The Hidden Curriculum—Faculty-Made Tests in Science

Part 1: Lower-Division Courses

INNOVATIONS IN SCIENCE EDUCATION AND TECHNOLOGY

Series Editor:
Karen C. Cohen, Harvard University, Cambridge, Massachusetts

The Hidden Curriculum—Faculty-Made Tests in Science
Part 1: Lower-Division Courses
Part 2: Upper-Division Courses
Sheila Tobias and Jacqueline Raphael

Internet Links for Science Education: Student–Scientist Partnerships
Edited by Karen C. Cohen

Web-Teaching: A Guide to Designing Interactive Teaching for the World Wide Web
David W. Brooks

A Continuation Order Plan is available for this series. A continuation order will bring delivery of each new volume immediately upon publication. Volumes are billed only upon actual shipment. For further information please contact the publisher.

The Hidden Curriculum—Faculty-Made Tests in Science

Part 1: Lower-Division Courses

Sheila Tobias and
Jacqueline Raphael
California State University, Dominguez Hills
Carson, California

PLENUM PRESS • NEW YORK AND LONDON

Library of Congress Cataloging in Publication Data

Tobias, Sheila.
The hidden curriculum—faculty-made tests in science / Sheila Tobias and Jacqueline Raphael.
p. cm.—(Innovations in science education and technology)
Includes bibliographical references and index.
Contents: pt. 1. Lower-division courses.
ISBN 0-306-45580-3
1. Science—Study and teaching (Higher)—United States—Examinations, questions, etc. 2. Educational tests and measurements—United States. 3. Grading and marking (Students)—United States. 4. Educational change—United States. 5. College students—United States. I. Raphael, Jacqueline. II. Title. III. Series.
Q183.3.A1T58 1997
507'.1'1—dc21 97-1865
CIP

ISBN 0-306-45580-3

A Division of Plenum Publishing Corporation
233 Spring Street, New York, N.Y. 10013

http://www.plenum.com

10 9 8 7 6 5 4 3 2 1

Printed in the United States of America

PREFACE TO THE SERIES

The mandate to expand and improve science education for the 21st century is global and strong. Implementing these changes, however, is very complicated given that science education is dynamic, continually incorporating new ideas, practices, and procedures. Lacking clear paths for improvement, we can and should learn from the results of all types of science education, traditional as well as experimental. Thus, successful reform of science education requires careful orchestration of a number of factors. Technological developments, organizational issues, and teacher preparation and enhancement, as well as advances in the scientific disciplines themselves, must all be taken into account. The current prospects look bright given national reform movements such as the National Academy of Science's "Standards for Science Education" and the American Association for the Advancement of Science's "Benchmarks"; the backing of science education leadership; and recent developments, including the Internet and new educational software. Further, we have a world-wide citizenry more alert to the need for quality science education for all students, not just those who will become scientists. If we can isolate and combine such factors appropriately, we will have levers for science education reform. The books in this series deal in depth with these factors, these potential levers for science education reform.

In 1992, a multidisciplinary forum was launched for sharing the perspectives and research findings of the widest possible community of people involved in addressing the challenge. All who had something to share regarding impacts on science education were invited to contribute. This forum was the *Journal of Science Education and Technology*. Since the inception of the journal, many articles have highlighted relevant themes and topics: the role and importance of technology, organizational structure, human factors, legislation, philosophical and pedagogical movements, and advances in the scientific disciplines themselves. In addition, approaches to helping teachers learn about and use multimedia materials and the Internet have been reported. This series of volumes will treat in depth consistently recurring topics that can support and sustain the scientific education enterprise and be used to raise levels of scientific knowledge and involvement for all.

The first four volumes illustrate the variety and potential of these factors. *The Hidden Curriculum—Faculty-Made Tests in Science: Part 1, Lower Division Courses* and *The Hidden Curriculum—Faculty-Made Tests in Science: Part 2, Upper Division Courses* are premised on the belief that testing practices influence educational procedures and learning outcomes. Innovations in exam practices that assess scientific understanding in new and more appropriate ways should be shared with the widest possible audience. The research described and the resulting compendium of hundreds of contributed, annotated best exam practices in all science courses at the college level is a resource for every science educator and administrator.

Web-Teaching: A Guide to Designing Interactive Teaching for the World Wide Web aids instructors in developing and using interactive, multimedia educational materials on the World Wide Web. It also helps instructors organize and control these resources for their students' use. Not only do instructors learn how to improve their own materials and delivery, but they can access and make available Web-based information in a way their students can comprehend and master. Using the lever of instructional technology is an increasingly important part of science teaching; this book guides that process.

Finally, *Internet Links for Science Education: Student–Scientist Partnerships* illustrates the workings and effectiveness of this new paradigm and growing force in science education. In these partnerships (SSPs), students help scientists answer questions that could never before be fully addressed due to the lack of a large number of strategically positioned observers. Students gather and analyze data in projects involving authentic and important scientific questions, and science teachers actively explain science to students and help scientists implement their research. Data gathering and sharing, the heart of effective SSPs, is possible and rapid with the help of the Internet and a variety of technologies—groupware, visualization, imaging, and others. Several representative SSPs are described in depth. Chapters on student data and the human and technological infrastructures required to support SSPs help readers understand the interplay of the several factors in this approach to improving science education K–12. The Appendix contains a useful annotated list of current projects with complete contact information. Readers of this book will come away with an understanding of these programs from multiple perspectives and will be encouraged to become involved in similar efforts.

It is our hope that each book in the series will be a resource for those who are part of the science reform effort.

Karen C. Cohen
Cambridge, Massachusetts

PREFACE TO PART 1

Every faculty member knows that exams drive student behavior. Testing and grading are part of every course but generally of lesser importance to faculty than course content and, more recently, instructional methods and pedagogy. But as issues of grade inflation, student attrition, accountability, and authentic assessment demand increasing attention, we see the beginnings of a reevaluation of the role of tests in the science curriculum and the introduction of some interesting innovations in science testing methods. If properly nurtured by the science community and by academic administrators, these changes could lead to substantial testing reform in the college science classroom. To that end, we submit this collection of testing innovations and hope that it becomes the first of a series of practitioner-generated, practitioner-circulated volumes about testing.

Two questions underlie this study: First, why is it so difficult to change tests and testing traditions in college-level science; and second, will the enormous efforts underway to reform pedagogy and curriculum in these disciplines have any lasting effect if testing does not have a prominent place on the reform agenda? Herbert Lin, a physics educator, calls examinations the "latent curriculum"[1]; Lucy Chester Jacobs and Clinton I. Chase, writing in a guidebook for faculty on developing and using exams, remind us that "tests send a message to students: 'Here are really essential things to learn and remember from this course.'"[2] Yet despite the significance of examinations, there is little sharing of exemplary practice among college faculty on this subject. This book is intended to fill that gap and to provide some new ways of thinking about and doing examinations in college science.

Even when the exams faculty use have serious flaws, change is hindered by institutional rules and cost considerations. Some barriers to change, however, are self-imposed. Science faculty are particularly sensitive to issues of fairness

[1] Herbert Lin, personal communication to the authors (1994).

[2] Lucy Chester Jacobs and Clinton I. Chase, *Developing and Using Tests Effectively: A Guide for Faculty* (San Francisco: Jossey-Bass, 1992), p. 3.

and objectivity. They remember, when they were students, being put off by "subjectivism" and capriciousness in course grading outside the sciences, and so fairness remains for them a high priority. Then there are those who feel that exams are not as urgent an arena of reform as, say, pedagogy, computer-assisted instruction, or curriculum. And as in all educational matters, inertia abounds. As one respondent to the solicitation that generated material for this book put it, "All systems of measurement are imperfect. Scientists know this." Said another: "If it's not broke, don't fix it."

In many instances, the nay-sayers have a point. Faculty may propose, but administrators will dispose. We asked science educators to address that issue in the Introduction to Part 2 of this collection, and their concerns are important. But most significant, the educators stressed how much exam reform is already taking place in college classrooms.

Our method for collecting exemplary practice was straightforward. We began by limiting our study to testing in the natural sciences, except where a submission in mathematics or engineering appeared to the authors to be deserving and applicable to science. A solicitation went out to a variety of mailing lists and through the 1994 Chautauqua Short-Course Catalog.[3] In addition, friends and colleagues distributed copies of the solicitation and advertised it in a variety of educational journals.

In soliciting "new theory, new practice," we asked college science educators to describe their innovations in terms of one or more of the following categories:

1. Exam design (including content in general and test-item construction specifically).
2. Exam format (such as length, open-ended, closed, multiple choice, verbal, pictorial, quantitative).
3. Exam environment (individual, group, in-class, take-home).
4. Grading practices (retesting, curve grading, opportunities to retake exams or to do extra work to raise grades).

We decided upon these categories because examinations are more than the sum of the questions or problem sets faculty invent to measure student mastery. Examinations embed assumptions about student learning, motivation, and attitude, as well as course goals and requirements. Although conscientious faculty generally spend long hours preparing the content of their exams, these other matters—exam format, exam environment, and grading practices—get less attention, despite the fact that they may be even more salient (in the minds of their students) than the content itself.[4] Many of the innovations described in this

[3] A copy of the solicitation is included in the Appendix.

[4] See "Listening to Students," p.15.

anthology are made up of a number of parts that can be modified and employed by faculty in different combinations, depending on their interests and needs.

Upon receiving a written or oral description of an examination innovation, we summarized or standardized the format of the summaries and went back to the innovator, by mail and telephone, for further detail and elaboration. Eventually, each submission included in this book was approved by the originator.

Throughout, we have had the cooperation of our contributors in conveying accurately to our readers what they are doing, what they are thinking about what they are doing, what is succeeding, and what is not. We also asked them to detail obstacles they and their colleagues are experiencing while attempting change in in-class examinations at colleges and universities. We are grateful for their enthusiasm for this project and for their willingness to share their ideas and thoughts. We also thank Francis Collea, and the California State University Foundation for its willingness to support this work.

CONTENTS

CHAPTER 1

LISTENING TO FACULTY

In reviewing the pattern of science education reform in the 1980s and 1990s, one is struck by how much attention has been paid to pedagogy and content in science and how little to changing the design, delivery, and grading of examinations. This, despite observations such as chemist Susan Nurrenbern's that "the last 30 years of chemistry can be characterized by the 100 most popular questions"[1]—and considerable documentation that students who successfully pass introductory college physics examinations have not really mastered Newton's three laws.[2] The same standard textbook problems continue to be rolled out year after year, "on a grand scale—by the hundreds of thousands," according to testing experts Ohmer Milton and James A. Eison, in all college subjects.[3] Especially in large universities, where additional grading assistance is considered an unaffordable luxury, these exams are in multiple-choice formats the faculty claim they are compelled to use even though they'd like to try something different.

Despite all odds, some faculty are experimenting, often alone and without collegial or administrative support. Their efforts, like ours in putting together this collection, are framed by two central questions: First, how can instructors make examinations more creative, more meaningful, and more useful as diagnostic and feedback mechanisms in the teaching of science? Second, will the current movement toward reform achieve any lasting effects if in-class examinations are *not* part of the agenda for change?

One difficulty, quite obviously, is class size. In a study of 1,100 college professors by T. L. Cross in 1990, faculty said class size was the most important

[1] S. C. Nurrenbern and M. Pickering, "Concept Learning vs. Problem Solving: Is There a Difference?" *Journal of Chemical Education* 64(1987): 508–510.

[2] D. Hestenes, M. Wells, and G. Swackhamer, "Force Concept Inventory," and D. Hestenes and M. Wells, "A Mechanics Baseline Test," published together in *The Physics Teacher* 30(March 1992): 141–158 and 159–166, respectively.

[3] Ohmer Milton and James A. Eison, *Better Course Examination Questions: Guidelines*, Publication of Learning Research Center, University of Tennessee, Knoxville, n.d.

factor in their decisions about the type of exam to use. And more than two-thirds of Cross's sample said the size of their classes prevented them from using the exam design they'd prefer under different circumstances.[4] Another barrier is time. Many faculty, says Peter Hilton, writing specifically about large lecture classes in mathematics, simply do not have the time to devise and experiment with testing ideas.[5]

A third reason exams are often less than effective in science (as in other disciplines), is that in-class testing is burdened with too many mutually contradictory functions. Agreeing on what we want to examine is clearly a prerequisite to designing new testing strategies and procedures. Consider the following "purposes" of examinations: (1) to rate students, absolutely and in terms of their peers (the main function according to many faculty today); (2) to rate both the professor and the course in the eyes of the department; (3) to motivate students to keep up with the work; (4) to provide feedback to the student as to how he/she is doing and in which subsets of the course material he/she has deficits; (5) to provide feedback to the instructor as to his/her success in meeting students' academic and other needs, and how and where to improve; (6) to provide a means of holding educators accountable; and (7) to enhance teaching and learning.

Scrutinizing the verbs embedded in these purposes, one finds that tests are supposed to diagnose, measure, compare, motivate, punish, scare, and create a competitive atmosphere, all the while enhancing teaching and learning. Intended to do too much, examinations end up achieving only some of their goals, and even those are often shortchanged.

Evidence that tests drive student behavior began to appear in journal articles as early as the 1930s. The student who told an interviewer in a 1935 study that in studying for multiple-choice, true–false, and completion problem examinations, he never bothered to do a "general review" of the material but only to "learn the facts or memorize the statements," is probably no different from today's college student, only more candid.[6]

Multiple-choice tests in particular have been faulted since they were first invented for obscuring how well students can make sense of altogether new material, handle anomalies and contradictions, draw or recognize pictorial representations of molecules and reactions, or construct coherent descriptions of physical or chemical events—all so-called higher-order thinking skills. In a lively introduction to the nonnumerical problem sets accompanying "Chem Zen," or "Voyages in Conceptual Chemistry" for nonmajors, Dudley

[4] T. L. Cross, "Testing in the College Classroom." Paper presented at the annual meeting of the American Educational Research Association. Boston, April 1990.

[5] Peter Hilton, "The Tyranny of Tests," *American Mathematical Monthly* 100, no. 4(April 1993): 368.

[6] M. Meyer as quoted in Milton and Eison, op. cit., p. 1.

Herschbach, Harvard chemist and Nobel Laureate, says the problem with typical textbook science problems is that they encourage three dangerous syndromes[7]:

1. The Plug-and-Chug syndrome—problems that encourage students to find the formulae and plug in the numbers to get the right answer.
2. The Just-the-Right-Data syndrome—problems that give students only the information needed to "solve" the problem so that they do not learn how to ask themselves what they need to know, or how to discern what information is significant.
3. The Don't-Know-How syndrome—problems that are so ordinary and safe that guessing and qualitative reasoning is effectively discouraged.

Not just the type of exam affects student behavior. Grading practices that test students individually and then apply a normal curve to the grade distribution provide information as to students' relative position in a particular class but discourage cooperative learning and studying. Exams that are designed so that the "average" student gets 60 percent correct may serve to differentiate the most able (the ostensible purpose of "curving") but will discourage the average student from persevering on an examination that is unduly challenging.

Faculty suffer right along with students. Many faculty acknowledge that standard exams do not give them the feedback they need to improve their teaching. As one of the respondents to our solicitation put it, when her students do not do well on an examination, she has no way of knowing whether she taught them poorly, whether her students were ill-prepared or not working hard enough, or whether her exams were simply too hard.

We offer here the proposition that the tests science faculty design and use are a significant part of the dynamic of a course and ought to be integral to its goals and curriculum. Since, as previously noted, examinations drive student behavior, imperfect testing strategies can sabotage the entire educational effort. Faculty have known this for a long time but as a group have not made finding a solution a high priority. As early as 1969, Eric Rogers, then a professor of physics at Princeton, described what he called "slippage" in the traditional physics course:

> Examinations tell [students] our real aims, at least so they believe. If we stress clear understanding and aim at a growing knowledge of physics, we may completely sabotage our teaching by a final examination that asks for numbers to be put in memorized formulas. However loud our sermons, however intriguing our experiments, students will judge [us and our course] by that exam—and so will next year's students who hear about it.[8]

---

[7] Dudley Herschbach, personal communication to the authors (1995).

[8] Eric M. Rogers, "Examinations: Powerful Agents for Good or Ill in Teaching," *American Journal of Physics* 37(1969): 956.

PREVIOUS EFFORTS AT REFORM

Thirty years ago, Leo Nedelsky, physicist at the University of Chicago, wrote *Science Teaching and Testing*, a book almost unique, then and now, first in addressing scientists (as opposed to educational researchers) on college and university faculty, and second by focusing on testing. His goal was to show teachers how much they could learn from their own testing. As he wrote, "Finding out what their students are learning is, for most teachers, the simplest road to course improvement."[9] Indeed, one way to help teachers teach is to teach them how to recognize and construct good tests, and how to interpret their results.

A good test, Nedelsky wrote, is one that gives a true picture of students' ability to use the information they have been exposed to in the course. But most science tests fall short by testing merely factual knowledge on the grounds that (1) factual knowledge easier to test; (2) only factual knowledge is taught, so it would be unfair to test anything else; (3) testing factual knowledge is a good test of everything else that matters; and (4) only factual knowledge can be accurately measured.

Yet, factual knowledge is not the only subject of science courses. On a scale of importance, Nedelsky places factual knowledge near the bottom. Higher up are "understanding" and "a measurable increase in a student's ability to learn on his or her own." According to Nedelsky, students possess knowledge if they can recall content; understanding if they can handle a situation that is novel to them; and ability to learn if (and only if) they can increase their knowledge and understanding without a teacher's help.[10]

Yet despite general agreement as to its worth, little of what Nedelsky recommended has been achieved. Like Nedelsky, testing experts Ohmer Milton and James A. Eison, writing in their guidelines for constructing examination questions, place *recognition* or *recall* of isolated facts at the bottom of a learning hierarchy, and *interpretation*, *assessment*, and *evaluation* at the top. Although in practice these levels overlap, the authors cite studies that indicate that the majority of test questions for undergraduates in all subjects favor isolated factual recall. In addition: "there is almost no emphasis on the higher order mental processing of comprehending, applying, and evaluating. Yet these abilities are the ones most faculty claim to develop in students."[11]

Indeed, Lucy Chester Jacobs and Clinton I. Chase, in *Developing and Using Tests Effectively: A Guide for Faculty*, single out college science exams

---

[9] Leo Nedelsky, *Science Teaching and Testing* (New York: Harcourt, Brace, and Word, 1965), p. xii

[10] Nedelsky, op. cit., pp. 20–21

[11] Milton and Eison, op. cit., p. 1.

in particular as flawed in the following ways: covering material not specifically discussed in class; not covering material the professor led students to believe was most important; asking ambiguous questions; and, in the way they are graded, providing limited feedback. Moreover, they found tests in science too short to cover the material taught, too few to measure accumulated achievement, and instructors unable or unwilling to convey to students the value of tests to their learning. These discontinuities, observed Chase and Jacobs, contribute to an adversarial relationship between instructor and students—especially around testing.[12]

Since Nedelsky's time, the development of qualitative and conceptual skills has become an important goal of classroom instruction. Lauren Resnick expressed the view that "educational practice should seek to embed efforts to teach cognitive skills in all of the traditional school disciplines."[13] As per the new science standards for K–12, classroom activities and exams are being designed that ask students to apply concepts to problems that more closely resemble real-world science, and educators are being urged to employ performance skills assessment in place of multiple-choice or other artificial measures. Absent motivational standards, what will motivate college-level science instructors to improve *their* in-class assessment techniques?

THE QUEST

Imagine a college or university science educator ready to change his/her in-class examinations. Where would that educator go to exchange ideas about constructing new test items or altogether new types of exams, or to get criticism and help from colleagues? One place is to the few books designed to help all teachers create better exams, books such as Jacobs and Chase's *Developing and Using Tests Effectively* or Milton and Eison's *Better Course Examination Questions: Guidelines*. But these how-to books are not discipline-specific and are therefore helpful only to a point. They will not provide the science instructor with test items, test formats, testing environments, and grading practices specific to science.

Another obvious but even less common strategy is to initiate departmentwide discussions of exams and grading practices, as chemistry faculty at Pennsylvania State University started doing in the 1993–94 school year. Department members took a critical look at the assumptions and practices underlying

---

[12] Lucy Chester Jacobs and Clinton I. Chase, *Developing and Using Tests Effectively: A Guide for Faculty* (San Francisco: Jossey-Bass, 1992).

[13] Lauren B. Resnick, *Education and Learning to Think* (Washington, DC: National Academy Press, 1987), pp. 34–35.

their course examinations, asking one another, what exams are we using? How are the results used? What are we *not* learning with the exams currently in use? What purpose do the exams serve for us as well as for our students?

It is just this kind of self-examination that researchers in the assessment movement say is needed to move from a "testing culture" to an "assessment culture," by which they mean a shift from ranking students by means of artifacts to documenting meaningful accomplishments.[14] Without a clear idea of what those accomplishments should look like—what Nedelsky calls the "goals of instruction"—the science exam may be nothing more than an exercise.

The kinds of test questions that stress comprehension and aim at a growing knowledge base are difficult and time-consuming to develop. Even more challenging is the need to think through: What are the goals of my course? How will I operationalize those goals in the tasks I assign to students? And do my exams measure how well I and my students have achieved those goals? With barely enough time to cover an ever-expanding syllabus, is it any wonder that science faculty try to get tests made, graded, and returned as quickly as possible?

Despite the barriers of class size and limited time, as this book documents, many science faculty are willing to change their exams. Other instructors want to explore new examination practices but need administrative support and the colleagueship of like-minded science faculty in their own and other institutions. But for now, those who are changing how they test students usually operate alone, without the benefit of such a forum.

NEW THINKING

Faculty represented in this collection found that thinking deeply about their teaching goals led them naturally to the design of better tests. In these efforts, some were helped by cognitive science, which has begun examining subject-specific heuristics in areas such as problem solving.[15] Research into how experts and novices approach physics problems differently, for example, may account for why thousands of U.S. undergraduates who can pass a course in mechanics cannot 6 months later answer basic questions about force, mass, and acceleration. With their "Force Concept Inventory," physics educators David Hestenes and Ibrahim Halloun showed that students can pass plug-and-chug in-class exami-

[14] Dennie Wolf, Janet, Bixby, John Glenn III, and Howard Gardner, "To Use Their Minds Well: Investigating New Forms of Student Assessment," in Gerald Grant, ed., *Review of Research in Education*, vol. 17 (Washington, DC: American Educational Research Association, 1991), pp. 31–74.

[15] Frederick Reif and Jill H. Larkin, "Cognition in Scientific and Everyday Domains: Comparisons and Learning Implications," *Journal of Research in Science Teaching* 28, 9(1991): 733–760.

nations without in any way eliminating previously held misconceptions about mechanics.[16] A growing concern among college chemists and physicists alike is that students are not developing conceptual understanding of basic principles, only algorithmic (numerical) proficiency. Andrea diSessa and Don Ploger put it this way: "[T]he central problem of science education is conceptual, not a matter of [teaching] efficient problem-solving strategies."[17]

In an effort to test whether algorithmic mastery of introductory chemistry corresponds to the kind of conceptual understanding diSessa and Ploger are writing about, Mary Nakhleh of Purdue persuaded her colleagues teaching all four general chemistry courses offered during a single semester at Purdue to add five matched sets of questions—each within a specific area of chemistry—to their final exams. One question of each set was phrased as an algorithmic problem-solving question, the other as a conceptual question that required understanding of the principles of the topic rather than mere mathematical reasoning. Three of the conceptual questions required that students present an interpretation of drawings, two a verbal response.[18]

Nakhleh found that 85 percent of the students (varying some by course) could supply the algorithmic solution; but only 49 percent could handle the conceptual question. The converse was not the case: Few students who failed the quantitative reasoning part of the question got the conceptual part right. Nakhleh's finding—that algorithmic facility does not guarantee conceptual understanding—corresponds to what has been documented in introductory physics.[19]

Herbert Lin, who studied physics education with Philip Morrison at MIT and with Lillian McDermott at the University of Washington, two giants in the area of physics education reform, achieved the same results by analyzing students' responses to what he called the "implicit" and "explicit" curricula of the introductory physics course he taught. Interviews with students during the course revealed that although the instructor was teaching physics, the students were taking a physics course that consisted not of the laws of motion, heat, and optics, but of ten problem sets, four lab reports, three 1-hour quizzes, and a final exam.[20] If the two course agendas did not compete for student time and attention, there would be no problem. However, from Lin's interviews emerged the obvi-

---

[16] Hestenes, Wells, and Swackhamer, op. cit.

[17] Andrea A. diSessa and Don Ploger, "Cognition and Science Education," in Audrey B. Champagne and Leslie E. Hornig, eds., *Students and Science Learning* (Washington, DC: American Association for the Advancement of Science, 1987), p. 21.

[18] Mary B. Nakhleh, "Are Our Students' Conceptual Thinkers or Algorithmic Problem Solvers: Identifying Conceptual Students in General Chemistry, *Journal of Chemical Education* 70, no. 1 (January 1993): 52–56.

[19] Hestenes and Halloun, op. cit.

[20] H. Lin, "Learning Physics vs. Passing Courses," *The Physics Teacher* (March 1982): 151–157.

ous: Physics is one of four or five courses that students handle by "shifting urgencies." Since they are graded only on the tasks assigned, the "implicit curriculum" has the stronger influence on their day-to-day behavior. For Lin, plug-and-chug—the bane of physics instruction—is the result not of student indifference to higher principles and higher-order thinking, but of *course demands*.[21]

Lin began offering his students more sophisticated techniques, with longer-term payoffs. At first, his students complained: "Your techniques take too much time. We have to do the problem set the night before it's due." Lin uncovered a typical survival strategy: "to force what I know into the problem rather than use the problem to determine what I need to know." Lin's solution (only successful when he gained complete control over the examinations in all sections of the course) was to insist that his students follow a set of specific guidelines meant to draw their attention to the *coherence* of their answers every time they solved a problem.

In grading, Lin evaluated student work first on content alone, then on coherence alone, and then multiplied the two grades so as to penalize omissions. Students resisted at first, but once the scheme was in place for all homework assignments and examinations, it succeeded in changing student study habits.

NEW PRACTICES

There are many imaginative ideas about testing in science that, because they are not evaluated in any systematic manner, or because they are too humbly regarded by their creators as little more than slight modifications of traditional testing tools, do not find their way into the literature on testing. But when specifically asked, science faculty have much to report about changing their examinations. Our study documents that college faculty today are altering their exams more systematically than they realize, and with more powerful effects.

For example, Timothy Slater, then at the University of South Carolina, now at Montana State University, decided to try portfolios in his physics and astronomy classes after reflecting on his own experiences before and during his teaching career:

> Throughout my tenure as a college student... instruction has always been lecture, test, lecture, final exam. [I was looking for] an "alternative" to the monotonous, traditional lecture-testing cycle.... In science, several brief observations are rarely enough for the scientific community to reach a consensus about nature—it seems rather odd that these same scientists would use brief glimpses of student performance to evaluate student understanding of the course learning targets.[22]

---

[21] Ibid.

Slater eventually selected portfolio assessment as more revealing and more comprehensive than traditional testing techniques. Test anxiety is minimized, and his students perform as well on standard exams as those who take traditional exams throughout the course.

At the University of Wisconsin, chemistry professor Arthur B. Ellis wished to replace simple one-step questions with the kinds of multistep problems more likely to be found in real-world chemistry. But he worried that his students, failing to recall the right formula or lacking the information needed for the first step, would panic and not do well. So, he invented a way to allow students to "purchase" clues from a "test insurance page" while taking an exam. "The basic idea of Test Insurance (TI) is that if a student is stuck, (s)he may buy a clue to an essay question or buy the answer to the first and/or second parts of a multi-step question for some fraction of the question's total worth. In our test, this "point premium" amounted to 20 percent (one clue or answer) or 40 percent (two clues or answers) of the question's value."[23]

Ellis employs "scratch-off" paper, the kind used in instant winner lotteries. Each student receives a "Test Insurance Page" at the start of an examination, in which hints are identified by problem number, page number, and point deduction, but unless "used" by the student, are concealed. Removal of the coating (by means of a coin) is irreversible and clearly indicates that a clue or answer has been seen. About two-thirds of the students in his class use at least one of the six clues available for each examination taken.

Also significant in chemistry education is the American Chemical Society's development, in 1994, of a conceptual examination for general chemistry, described on pages 73–74 of this book.

Other instructors are making efforts to change the exam environment—that is, to break the competitive aspect of examinations in science. Typical of efforts to employ group exams is an innovation begun 4 years ago in an introductory earth science course at California State University at Chico. Instructor Ann Bykerk-Kauffman first introduces the group format on a midterm exam and then allows students to choose for their final exam whether to take the test individually or in a group. (The group test is "harder," meaning that it contains more complex questions.) Members of the group may discuss each question among themselves but must then answer individually, rationalizing both their own answers and any divergence within the group. Using the group format, the instructor is able to give multiple-choice questions requiring higher-order thinking skills.[24]

---

[22] Timothy Slater, personal communication to the authors (1995).

[23] Arthur B. Ellis, "Test Insurance: A Method That May Enhance Learining and Reduce Anxiety in Introductory Science Examinations," *Journal of Chemical Education* 70, no. 9(September 1993): 769.

[24] Personal communication from Ann Bykerk-Kauffman (1994).

From a questionnaire attached to the final, Bykerk-Kauffman has been collecting student feedback as to the fairness and usefulness of the new format. Even when questions are difficult, she finds, students seem to enjoy themselves, as does she. "I thoroughly enjoy seeing my students spend a full 50 minutes vigorously discussing and debating geology. In most cases, the students completely reject most of the detracter answers. So the groups are effective in eliminating the dumbest answers."

Even more daring is chemistry professor Ralph Dougherty's replacement of the exam-generated final grade in the course with an "Optional Grade-Performance Contract." The course is organic chemistry, and the experiment has been running several years.

Dougherty's initial premise was that if his organic chemistry students were to do all the work he recommends that they do, they would pass his exams. To demonstrate this, the instructor offers an "Optional Grade-Performance Contract" in which the work he thinks is needed to master introductory organic chemistry is laid out in detail. To make his point, Dougherty guarantees students a passing grade—whatever their numerical scores along the way or their numerical average at the end—if they fulfill the requirements of the contract. If, at any time, a student fails to meet the conditions of the contract, the contract is voided and the students is back on his/her own.[25]

The "minimal requirements" are not minimal. They include reading all appropriate text material prior to the lectures, attending all lectures and taking comprehensive notes, attending one recitation per week, and transcribing all lecture notes in ink from lectures and recitation sessions, supplemented with complete examples.[26]

The bound notebook is the linchpin of the contract. If a student under contract fails an exam during the semester, he/she must see the instructor with notebook in hand. If the notebook is incomplete or inadequate (or if the student does not come in at all within 48 hours), the contract is nullified and the student's earned grade-point average in the course will be the final grade.

With the new system, Dougherty intends to convey to his students just how much time is needed to learn organic chemistry. As part of their contract, they are to log their hours spent and work done, with a minimum of 9 hours of work per week exclusive of lectures and recitations, with no more than 2 hours counting for any one day. In addition, students are to (1) work all assigned problems and cooperate with a group to turn them in for credit; (2) work with a group on all quizzes and pretests; (3) review all notes for the course not less than once a week; and (4) note the review in the log.

---

[25] Ralph Dougherty, personal communication to the authors (1993).

[26] R. Dougherty, S. Tobias, and J. Raphael, "The Contract Alternative: An Experiment in Teaching and Assessment in Undergraduate Science," *AAHE Bulletin* (February 1994): 3–6.

In the first year in which the contract was offered, Dougherty's failure rate (Ds, Fs, and Ws) dropped from the typical 50 percent, where it had been for the past 10 years, to 19 percent. But more importantly, says the instructor, the contract arrangement forces him to interact with the very students he was missing in the past: those who do poorly but are hardworking.

Dougherty's low failure rate, of course, could be attributed to the fact that contracting students must pass. To persuade his colleagues that his students are actually achieving as much as theirs, Dougherty compares this class with a traditionally graded organic chemistry class by means of the American Chemical Society's standardized organic chemistry exam. His contract students do just as well or better on average than the others, and the number of As and Bs is higher in the class in which the contract was offered than in the control group.[27]

COMPUTER-GENERATED EXAMS AND SCORING SYSTEMS

Even though computer-based assessment in college-level science has been associated mainly with scoring systems rather than exam enhancement, we would be remiss in describing new theory and new practice in science examinations without mentioning the additional possibilities presented by computer-based assessment. Our solicitation brought in only a few examples , but there are certain truths that education reformers ignore at their peril: first, that technology is already deeply entwined with large-scale, out-of-classroom testing. Indeed, it has been asserted that without technology, large-scale testing, with its requirements for test-item analysis, interrater reliability, and validity, would not have been possible.

Second, within classrooms, computerized test-item generation and scoring are already widely used in college-level science. Test banks on disk are frequently provided by textbook publishers. Because of class size and the traditional fact-based curricula, machine grading has become standard fare, so much so that even in laboratory-based courses, instructors employ multiple-choice, true–false, and computational questions.

Past disappointments notwithstanding, we are on the brink of some real breakthroughs in the quality of problems that the computer can generate and score, including the handling of "constructed response" (as against passive recognition) answers, open-ended (as against single-answer) questions, and items that test students' ability to recognize, model, and manipulate symbols and graphic images.[28]

---

[27] Ibid.

[28] See the special issue on computer-based science assessment, *Journal of Science Education and Technology* 4, no. 1 (New York: Plenum Press, March 1995).

As concern for school-to-work readiness increases, instructors will be pressured to test how well students perform in the computer environment in which they will work. Team-based, real-time simulations are common in industry. Although, as many have noted, technology penetration in the workplace far outdistances technology penetration in the schools,[29] "performance-based assessment" in science may soon come to mean ability to perform in a worklike environment, which includes computers. Equity goals, too, may require that performance other than paper-and-pencil composition be available to students having language background incommensurate with their scientific understanding.

Already, as some of the contributors to this anthology indicate, instructors are finding students' e-mail responses to open-ended test questions less tortured and therefore more representative of their true level of understanding than handwritten answers in class. Others are finding that the computer helps in weighting quality of answers or number of retakes, allowing students to "make up" lost or missed points, even to get partial credit. Another potential use for computer technology in testing involves the tracking (and even the analyzing and eventual tutoring) of student problem-solving strategies or protocols. As described by David Kumar and Stanley Helgeson, in a recent special issue of the *Journal of Science Education and Technology* devoted to computer-based assessment,

> Classic problem solving, such as calculation of molarity of a solution from the amount of solute and solvent, involves multiple steps and higher-order mental behaviors such as application, comprehension, etc. In large-scale testing, it is customary to grade student performance in such situations by merely grading the final answer (a product).... Computers offer a hypertext/hypermedia environment for developing systems for assessing nonlinear problem-solving tasks.[30]

Another arena in which the computer offers new possibilities is "adaptive testing," in which test time is shortened by administering to each test-taker only those questions tailored to his/her competence level. This technique is already being used in achievement tests administered by Educational Testing Service.[31]

Whether science instructors are ready for computer-based testing may be a moot question as software becomes an essential part of the teaching of science. Enormous progress has been made in instructional software, especially in chemistry, where interactive spreadsheets, simulations, and other sophisticated software are now used. We must assume that tests will continue to reflect

---

[29] Hessie Taft and Mark Singley, "Open-Ended Approaches to Science Assessment," *Journal of Science Education and Technology* 4, no. 1 (New York: Plenum Press, March 1995), p. 19.

[30] David D. Kumar and Stanley L. Helgeson, "Trends in Computer Applications in Science Assessment," *Journal of Science Education and Technology* 4, no. 1 (March 1995), p. 33.

[31] Taft and Singley, op. cit., p. 17.

instructional materials, so that as technology becomes more and more a part of the learning process, it will seem only natural that students take their tests on computers as well. Just as the graphing calculator has changed the kinds of questions mathematics teachers are asking in elementary college courses, computer-generated data analysis, dynamic graphs, and computer-generated animation will suggest inquiry-based science questions that evaluate the student's ability to work out answers on the computer.

Test-taking from this perspective is but a subset of the larger task of communicating scientific insight and information. If, as John Moore, who publishes *Journal of Chemical Education: Software*, predicts, professionals will be communicating scientific information differently in the future, "[employing] graphs and animations as routinely as [they] now employ text, figures, and tables,"[32] then communication from student to instructor may have to change, too. More important is to convey to students the iterative nature of scientific inquiry along with the importance of criticism and self-criticism—teaching challenges that are and will doubtless remain technology independent.

CONCLUSION

What David Hestenes, Mary Nakhleh, Herb Lin, Timothy Slater, Arthur Ellis, A. Bykork-Kaufmann, Ralph Dougherty, and others are trying to do is to sort out the feedback and teaching functions from the grading and comparing functions of science testing and then make these functions coalesce in as fluid and effective a manner as possible. College instructors have far more freedom to experiment with examination types and formats than do pre-college teachers faced with the Advanced Placement Test, the Student Achievement Test, and other standardized measures of achievement. That is why there will probably be no lasting change in testing in college and university science until and unless the Graduate Record Exam and other standardized preprofessional examinations are changed as well.

Alternate modes of examinations—as several of the contributors to this anthology demonstrate—need not be more expensive in the long run. But creating a new assessment mode and training faculty to operate happily in the new mode may well be. The innovations contained in our book bear witness to how much can be done without additional cost. But even where additional resources are required, appropriately measured, the benefits to students ought to substantially exceed the costs.

---

[32] John Moore, Editorial, *Journal of Chemical Education: Software*, 4C, no. 1(1992): 10.

CHAPTER 2

LISTENING TO STUDENTS

What do our tests tell our students about themselves? And about science? In an effort to explore student recollections of examinations in college-level science, the authors conducted a series of short focus groups with thirteen undergraduate and graduate students. All had taken more than 2 years of science or engineering at college (most at the University of Arizona) and had done well. A small number reported on experiences elsewhere, including one who attended a liberal arts college (Occidental), and two who attended a community college before transferring to the University.

While not comprehensive, this purposive sampling of student attitudes balances the collection of narratives that follows, written by and about *professors'* efforts to deal with student objections and their own misgivings about traditional examination practices.

EXPECTATIONS AND DIRECTION

The students we queried tended to be better-than-average test-takers, but they were mindful that instructors serve many students not as intrinsically motivated as they are, for whom tests serve as goads. Still, some of our older students were unprepared for the younger students' lack of interest in courses. "If you tell me what to memorize and study, I'll get a C and be happy," is a younger student's attitude that frustrates Sara. The older students are even more surprised that faculty treat college students "like children," giving grades for attendance and participation, grading homework solely to make sure it gets done, and hinting to students about what will be on the exam.

From our discussions, we sensed that although our better students appreciate being told what is expected of them, at the same time they are frustrated when exams so closely follow lectures that reading the text is redundant. That is, they are unhappy with tests that are too predictable.

EXAM FORMAT

Many of the focus-group participants prefer short-answer or essay questions over multiple choice in science, but for different reasons. The most surprising reason was given by Joanna, who finds grading "less fair" with multiple choice because students who "know more than what the professor thinks they know" find nuances in ostensibly straightforward questions and are distracted by the exceptions. In cases in which a question doesn't yield one unique correct answer, multiple choice "doesn't give you credit for what you know," says Joanna.

A distrust stems from the "trickery" students feel is at work in the instructor's construction of multiple-choice questions. "Whenever I'm taking multiple-choice tests," says one of our participants, "I always get the feeling someone is trying to trick me. That's how a multiple-choice test can be difficult: if it's designed to methodically trick the student into thinking something different." Another student feels the class ecology changes entirely with multiple-choice exams. What frustrates her is that, instead of being taught, she feels someone, such as the instructor, is whispering in her ear, "Now I'm going to catch you." The student is referring, of course, to the distracters. One of our contributing faculty members reviewing the research literature for new testing ideas found evidence that incorrect multiple-choice items actually reinforce incorrect answers to some degree.

Multiple-choice testing is actually unique to this country and not employed everywhere among U.S. institutions with the same frequency. Some students we talked to were rarely given multiple-choice tests in science, whereas others had them often. The decision appears to rest with school and professor. In a special inorganic class offered through Harvard Extension for students preparing for medical school, quantitative questions were the rule, with lots of partial-credit grading. But at Syracuse University, in a more traditional program, the same student reported that her biology professors preferred tricky multiple-choice exams. And later, brushing up on mathematics at a community college in Tucson, her exams were almost all multiple choice.

Lynn, with a B.S. in astronomy and one semester away from another B.S. in physics, finds that exam formats also vary with course level. In introductory astronomy courses at the University of Arizona, her exams were mostly multiple choice, calling for memorization, as she saw it, but not so once she got to higher-level astronomy and physics courses.

Multiple-choice and true–false exams tend to be more predictable than others. Students in our focus groups say that if they can "figure out" what the instructor really cares about, they can easily predict the multiple-choice questions that will be on an exam. Certainly, students need to be learning what faculty feel are the most important concepts of the course. But our students reveal that these exams are oversimplified *because* of the multiple-choice format. Once a clever student recognizes the teacher's "pet concepts," he/she can predict what kinds of multiple-choice questions will be on the exam and not have to study. "I got an 'A' in [such a] course and I learned nothing." In the worst cases, multiple-choice tests are recycled year after year, says one of our students. These "lazy-man" tests are worthless, according to our participants, especially when students circulate (or sell) them to next year's students.

Mature students, such as Brian and Joanna, remain frustrated by multiple-choice exams, not because of their trickery, but because they want the instructor to know how well they understand the material. The problem, they explain, is that many questions are borderline: "Depending on your interpretation, different answers could be given." Often, they resolve their distress by writing gratuitous explanations for their answers under the question, such as, "I chose B because..."[1]

David particularly dislikes multiple-choice questions that ask students to identify more than one correct answer out of five. These questions are "more like legal problems" to him, too nitpicky and, again, too tricky. Still others fault multiple-choice questions for testing mere memorization: "You can learn all the facts you want, but it doesn't mean you understand anything." Dave and Sara, also older students returning to science courses in pursuit of new careers, note that multiple-choice questions test students' ability to *recognize* right answers, a different mental task than that required to *generate* an answer to an open-ended or a calculational problem, or to *recollect* an answer on their own. "I'm much more inclined to study hard for a test in which I'm going to be required to recollect something rather than merely to recognize it," another science student puts it.

Having the information laid out in multiple-choice exams discourages those who pick up science quickly from going over the material at home and synthesizing information from multiple sources. In summary, we found that the better and more motivated the student, the higher the level of dissatisfaction with multiple-choice exams.

COMPETITION IN GRADING PRACTICES

One practice that some feel contributes to competition is grading on a curve. As a student in an earlier study observed:

> My [physics] class is full of intellectual warriors who will some day hold jobs in technologically-based companies where they will be assigned to teams or groups in order to collectively work on projects. [But] these people will have had no training in working collectively. In fact, their experience will have taught them to

[1] See Michael Zeilik's entry, p. 190, in which students are allowed to challenge any one of the questions on his exam.

> fear cooperation, and that another person's intellectual achievement will be detrimental to their own.[2]

When exams are too hard, competition and, for some, anxiety soar. Curves turn friends into enemies.

Mary, who holds degrees in humanities and theology, and is working toward a degree in higher education, was able to compare science exams to those in other disciplines. In her science classes, exams defined her study strategies more than in any other subject area. "In science, your grade depends almost entirely on exams," she says, "and if the grade is important to you for graduate school or other professional plans, then you are forced to put the grade first, before the learning."

Anne, who recently got her B.A. in biology, agrees that tests carry too much weight in science courses: "The problem with not testing well [in science] is that your mentors and all the faculty you're involved with don't think of you as a good science student, whatever your other attributes."

Competition is compounded by the relative difficulty of earning an honor grade in science in comparison to other college subjects. Some of the university professors our students encountered baldly stated to them that there was no point trying to get an A in their course: "No one does. Expect a B," students were told. Such attitudes have different effects on student motivation. Mary felt driven to prove her abilities; Lisa, a Hispanic engineering student, was made to feel "lucky" (not deserving), even when she did well on an exam.

Not all faculty intend to have students compete. But the "testing culture" in science courses forces comparisons. Anne describes herself as getting "slammed" with a bad grade on a difficult exam and finding it "hard" to separate herself from her grade, which made it a "sour" experience. Jeff, a wildlife biology major, says of a similar experience, "You end up hating the subject, hating the professor, and everything else about the course."

There are other negative effects of grading on a curve, say our participants:

1. The professor isn't responsible for tests that are too difficult because he/she can curve grades up. Our participants felt professors should take responsibility for the level of difficulty of exams.

2. Getting a 40 percent or 60 percent on an exam sends a negative message to students about their potential for success in science—and probably about science itself. Curving is "okay by me," says David, but leaving an exam knowing he couldn't possibly have gotten more than 60–70 percent of the problems correct made him uncomfortable.

---

[2] Eric Schocket, as quoted in Sheila Tobias, *They're Not Dumb, They're Different: Stalking the Second Tier* (Tucson, AZ: Research Corporation, 1990), p. 24.

3. Curving reinforces the higher value of *relative standing*, rather than knowledge gained or hard work, as a student in electrical engineering complained. One of our students also wondered whether it was ever fair, given the range of students in her electrical engineering classes (including some who have worked in engineering for years), to judge all these students' performances comparatively.

Joanna and others note that curving often is used to compensate for exams that are poorly designed, too simple, or too difficult. But she critiques its very *premise*: If the whole class is lazy and hasn't learned the material, why should their exam scores be curved?

Sara, on the other hand, liked curving while she was in college because she benefited from it. She noted that, after a particularly difficult exam, some faculty seemed to take cues from students' raw scores and made improvements in their teaching and in designing the next exam.

OTHER GRADING PRACTICES

David and Sara wonder whether grading on a curve actually obscures achievement by averaging, or leveling, students' accomplishments: "A good indication that you are learning is that you get progressively better grades on tests," says David. Yet, the averaging of grades throughout the semester penalizes the improving student.

Being graded on improvement, even in a small way, struck many of our respondents as fair and reasonable.[3] And the idea of mastery learning led to a good discussion. Anne recalled an extremely popular math teacher at the local community college who allowed students to repeat an exam up to five times. Yet these grading practices were atypical in our participants' experience.

Students of science acknowledge that courses using more open-ended exam questions require faculty to employ different grading practices from what they and their students are used to. One participant in our group remembered having a teaching assistant (TA) who circled key words in her students' responses to essay questions. If a student missed some key words, even a good essay earned her a bad grade—a kind of lazy man's grading.

Some of our engineering and chemistry students recalled, with more than a little horror, the "third significant digit" professor, who graded an entire problem wrong if it had the wrong third significant digit. "All this hard work to no end," complained one student.

---

[3] A number of examples follow in our collection for ways of measuring improvement.

Joanna brought up another important issue: the point value given for different questions, particularly the undervaluing of essay questions. "When an essay question on an exam is worth 20 points and there are 40 multiple-choice questions valued at 20 points as well, it's frustrating, because you want to give the essay question a lot more time," she says. But what she considers to be the "silly questions," such as matching pairs, are sometimes valued more.

COOPERATIVE LEARNING AND ASSESSMENT

Because cooperative learning and assessment is of interest to educators today, we asked our participants if they'd encountered these techniques in science courses and how these methods made them feel about the work. David likes to work in pairs in his lab classes, but he does not like taking quizzes cooperatively. As a teacher, he has given group tests in his Spanish classes and has come to doubt that all students participate equally in such exams. As with cooperative learning, there are "leaders and followers," so he isn't sure if all students really learn this way either. Because some people (like himself) prefer to work alone, David chooses to give his students a range of options. Another student describes herself as a "lone wolf when it comes to studying." And when an instructor gives a take-home, she says this penalizes her: even if they're told not to, she says, students compare answers.

Sara says that "cooperative learning is valuable so long as there's some way to assure equal contribution from everyone in the group." She could recall one good cooperative exam in sociology. Six essays for each unit were divided among groups of students, but each student had to write his/her own individual answer for the exam.

TIMED EXAMS

"We may have certain time constraints in our lives," says Burton, "but there have been dozens of tests on which I didn't do well because I just didn't have enough time." Some students have other horror stories to tell. Gillian was given 50 minutes to complete a 10-page essay exam in physiology. When students complained to the professor about the time pressure, they were told that if they really knew the material, they would have been able to "spit it out." "We were tested on how quickly we could write," she recalls, still bitter. This is an extreme case, but subtler insights were given by our participants. About all her exams, Mary says, "The more I know, the less time I have," meaning that when she really knew what she was writing about, the time constraints were more oppressive.

Most of our participants saw timed testing as an unnecessary stress that made them lose their train of thought. "I may know the material, but just watching that clock tick away is enough to make me draw a blank," says Anne. She feels better when the professor helps pace the students, telling them, at regular intervals, where they should be to get the exam done. Lynn points out that checking answers, something scientists do constantly, is extremely difficult when trying to get through an exam under time constraints. Students find themselves having to weigh such options as leaving the last question blank and checking their answers, or making a stab at the last question, anxious and under pressure.

VALIDITY

The students we queried challenged the validity of their exams when they noted important science skills, such as lab skills, entirely absent from their course examinations—even when the laboratory is an integral component of the course. "If the labs are so important, why isn't any of *that* content included on exams?" asked one student. Estimating skills are valued by their instructors but are rarely tested, nor are the "hardest and most useful thinking skills" these students remember getting from their science classes.

A number of our participants found answers on science tests to be either right or wrong ("black" or "white," as they put it), and they didn't feel this reflected the real nature of science. For Sara, returning to basic science classes in her 40s, "If you're trying to describe a system or a process, it's not just explaining a small part, but understanding the whole," which is why she prefers diagrams, essay, and short-answer questions to multiple-choice and true–false examinations.

GETTING HELP

Students have mixed feelings about seeking help from professors during classes or office hours. Some fear the professor will remember, when grading exams, if the student even once appeared unprepared or incompetent. "I don't feel comfortable going in to ask questions during office hours," says one respondent. "The professor might discover I haven't studied enough." Others don't feel their professors are accessible or objective enough during these meetings and that they later punish them for their coming for help. Although these perceptions cannot be validated, it might alleviate students' misgivings if professors made a point of explaining the relationship between seeking help and getting good grades in their courses.

Sometimes Lisa doesn't feel comfortable seeing a professor during office hours because she feels that she will be negatively prejudged as a female in her male-dominated electrical engineering major. "It's like an interview," she complains, and some professors have preconceived notions or personal prejudices.

What she is getting at is the conflict of having the same instructor serve as both the primary resource for extra help and the arbiter of grades.

BEST AND WORST EXAMS

In each of our focus groups, we asked students to recall the "best" and "worst" exams in their science courses. Sometimes their answers revealed their own evolution as science students. Anne's worst exam turned out to be a conceptually oriented test given at the university to which she transferred after having begun her work in science at a community college in the same town. She "froze," being unprepared for the deeper, more comprehensive questions. Nor had the homework assignments prepared her for an exam that provided an experimental scenario including results, which she was asked, on the spot and under time pressure, to analyze. Her best exam was a similar one that she took 2 years later, "when I was able to think one of those conceptual tests through." When she and the test fit together in her evolution as a thinker, Anne did her best. But these "fits" were few and far between.

Jeff's best and worst exams were in the same biology class, team-taught by three instructors. The worst, as he remembers it, was "an extremely tough multiple-choice test filled with holes and many kinds of problems." The class as a whole did poorly. "After I took the test I felt like I didn't know anything," he says. The best test was a few months later when there was less material covered and the test was more straightforward.

Burton's best exam took place in his high school Advanced Placement class, but he admits he has trouble distinguishing a well-formulated exam from one that he does well on: "In my mind, a great test is a test I do well on and am well prepared for, one where I don't feel like I'm being tricked." His worst exam took place in an alternative calculus course taught at a small liberal arts college. He remembers the test as "terrible" because it didn't correspond to the textbook. "It was 20 pages long, with a different problem on every page. It was very intimidating in that it made me 'do this!' and 'do that!' I felt like a machine."

Tonya's worst exam was in animal science. "The exam was almost all multiple choice, the kind where an answer could be a, b, or a and b, and so forth." On the one hand, she felt, "You definitely had to know the stuff to do well on one of these," but she regretted that it was so "tough." Another bad experience for her occurred in calculus, where test problems on the exam were harder than homework problems. In this instance, the instructor did not curve the grades, even though the highest grade was 79. She and about half the class failed and had to retake the course. Her best exam was a hands-on assessment in which her class visited centers at a farm and had to demonstrate what they knew in context.

David's best testing experience was in a chemistry course at a community college in which the tests were open book, open notes, and untimed. The tests were difficult—and the challenge was part of David's learning experience—but without time pressure, he felt no testing stress and was able to demonstrate that he had learned a lot. The instructor employed a variety of formats and offered a mix of problem types on each exam. His worst testing experience was in a biology course in which the instructor graded essay questions on criteria that were new for David: writing just the right amount and focusing on the question asked. David felt the grading of essay questions in the course was unfair.

Joanne's worst exams were true–false tests given in an introductory astronomy course. The questions were hard for her because she thought too much about the statements and ended up getting answers wrong. She had to approach the professor after almost every exam with three or four questions she found ambiguous. "I had a different angle on the questions, and the phrasing made them unclear," she says. In this instance, her professor was responsive and, as she remembers it, it was a "good experience because I remember a lot as a result of arguing with my professor."

Her best exams were in a biology course, even though the questions were always difficult. Although she generally likes and does well on essay questions, it wasn't possible to "b. s." on these exams. "You had to *know* the material," she says. She went into the course not expecting to like it, but ended up loving the course and choosing this particular subfield for her major—mainly because she got so much out of the tests.

Sara's best exam was in physiology and was a "good integrative exam." On the final for the course, the essay questions made her "pull together" all the different systems (in the body) she had learned about in separate sections of the course. During the final, she saw how much she now understood about the way the body worked. She remembers, though, that throughout the semester, the instructor focused on integration and synthesis, using body functions to which students could relate to help them understand the material. The instructor was particularly good at verbal communication and synthesis. Sara's worst exam was in biology at a community college. She had studied hard for the exam, but when most of the other students came in on exam day unprepared and complaining, the instructor decided, at the last minute, to make the test open-book, which frustrated Sara.

Mary's favorite science testing experiences came when the answers she was able to give "accurately reflected the amount of work I did for the course." This only happened when an exam consisted of four (or fewer) questions on which she could develop and elaborate her answers.

Brian's best exams involve open-ended questions. His worst exam was an oral exam in the classroom, in which the professor shot two narrow, pointed questions at each student. All the questions were different, which didn't seem fair to Brian, nor did this method appear valid, because of the limited sampling of material.

Joanna echoes this concern: "A good exam is going to cover all the material, not just focus on one or two things," she says. Her most difficult exam, but also one of her favorites, was in a course in human evolution. Students had to create their own evolutionary trees unlike any they'd learned.

What our participants liked least about their exams in science were timed tests, exams allowing no choice, and not being given formulas on the blackboard, in a handout, or in an open-notes method. As for grading practices, better students appreciate it when an instructor gives credit on an exam for an insight that he/she didn't think about him/herself.

CONCLUSION

"Tests should be most innovative at the introductory level," says one of our students, who finds this not to be the case. It is precisely in their first-year courses, our students think, that concept-oriented, higher-order thinking questions should be asked, to prepare students for these kinds of questions, which will most certainly appear on upper-division exams. After a year or more of multiple-choice, true–false, matched-pairs exams, our students found themselves both surprised and underprepared for higher-order thinking questions when they finally appeared.

Essay questions, in fact, were preferred by all of our introductory science students, as long as they weren't worth more than 25 to 30 percent of the entire exam grade. Students particularly liked getting at least one essay question in advance (when tests were timed) so they could do their "higher-order thinking" beforehand. For Sara, the "essay" could be oral. Her ideal assessment (common in university science in other countries) is to have an instructor ask her in a one-on-one interview: "What did you learn in this course?"

The ablest students in our focus group shared one vision that could easily be realized: Professors ought to offer a diversity of questions and of format because the students in their classes have different learning styles. Again and again, we heard these students say, "I wanted an exam to reflect the work I put in." Many faculty, however, appear to have different priorities. When they have to choose, they appear to our students to value *accuracy* and *reliability*, the assurance that the exam is constructed in such a way that *any* grader reading it would give it the same grade, even more than *validity*, the relevance of an examination to the content and purposes of the course.

HOW TO USE THIS BOOK

What follows is the heart of this book: the descriptions by faculty members of their in-class exam innovations generated by our solicitation. Because higher-education faculty organize themselves and their courses by discipline, their entries are organized by discipline and subdiscipline in alphabetical order by faculty members' names.

The Index is meant to help the reader locate exactly the innovations he/she is interested in—by type, by discipline, by size and nature of course (for majors, for nonmajors, and so forth.).

If this collection performs a useful function, it is our hope that others in the science education community will continue to collect and publish examples of not just what faculty members are trying to do in the way of innovative in-class examinations, but also what problems in testing they are trying to solve.

INDEX

BIOLOGY

CHEMISTRY

CHAPTER 3

BIOLOGY

ORAL CLASS QUIZ WITH QUESTIONS GENERATED BY STUDENTS IN SMALL GROUPS

Frank Baker, Department of Biology, Golden West College, 15744 Golden West Street, Huntington Beach, CA 92647-0592; TEL: (714) 895-8134; FAX: (714) 895-8989; E-MAIL: fbaker@cccd.edu.

Courses Taught

- Introduction to Human Anatomy and Physiology
- Introduction to Biology for Nonmajors
- Human Anatomy

Description of Examination Innovation

In his biology courses, Frank Baker uses an in-class exercise that is cooperative, student-centered, and makes test taking fun. Though it is a mock quiz, he combines a number of innovative aspects of testing into a useful in-class learning experience.

His class of 50–80 students divides into five-member groups. For 20 minutes, the groups write six multiple-choice quiz questions, which they submit to Baker. The instructor, with between 50–100 questions, chooses those with good distracters to stimulate discussion. (The instructor reserves the right to modify student questions.) Then, the quiz is administered *orally* to the entire class. Students discuss the question, then vote on the answer.

"The majority rules," says Baker. "If they're right, the whole class gets the point. If they're wrong, everyone misses." Discussion may continue after a vote if a student feels strongly about an answer.

In addition, everyone must contribute in some way by the end of the quiz or the whole class gets a zero.

Although the mock quiz is taken seriously, it doesn't jeopardize students' grades. It encourages students to help each other and contribute to productive "cross-classroom" dialogue, which becomes lively, requiring effective leadership from the instructor. Baker has never given a class a zero, although once, he put the final question on the board, told the class some students hadn't contributed, and warned them that they all faced a zero. Then he left the room and returned in 5 minutes. While he was gone, Baker's quiet students had been tutored. When he returned, they led discussion of the remaining question and earned the quiz points for the class.

The mock quiz is one of Baker's most popular classroom strategies. Students enjoy it, often to the point of competing for "air space" in the discussion. Most feel they learn more through class discussion.

CASE-SCENARIO PROBLEMS ON ESSAY AND MULTIPLE-CHOICE EXAMS, DIAGRAMMING REQUIRED ON EXAMS, UNGRADED SELF-ASSESSMENTS

Evelyn S. Becker, St. Louis College of Pharmacy, 4588 Parkview Place, St. Louis, MO 63110; TEL: (314) 367-8700; FAX: (314) 367-2784; E-MAIL: ebecker@SLCOP.STLCOP.edu.

Courses Taught

- Introductory Biology
- Biomedical Ethics

Description of Examination Innovation

Evelyn Becker uses a number of different test strategies including a "case-scenario" format for multiple-choice and essay questions. Although this kind of question is commonly used on ethics and biomedical exams, Becker has found the style works well for traditional biology classes as well.

"Even though this is an introductory biology class, I want my students to learn how to explain the science in terms a layperson could understand," says Becker. The case-scenario problem is well suited to this teaching goal because it requires that students justify their answers. A multiple-choice format is used to present a situation, with a follow-up question asking them to justify their answer by choosing from a series of options. For example, one of Becker's test questions states the following: "Dr. Aro Bick wants to test the effect of regular exercise on heart rate. Two hundred volunteers are recruited into the study." The first question asks students to identify the independent variable in this study. The choices are (1) heart rate, (2) exercise, (3) diet, or (4) genetic background. The next question asks students why they chose the answer above, with options such as "An independent variable is one that must be controlled."

Having students justify their answers gives Becker a chance to assess student problem-solving and get some "insight into persistent, subtle misconceptions," says Becker.

"I'm always trying to get the student to answer a question, and then ask another in which he/she has to explain the first answer," says Becker. "In the traditional multiple-choice format, that's not always possible."

Becker's essay questions are also based on case scenarios. Early in the introductory biology course, the student is asked to design an experiment to determine whether study time is directly proportional to the grade students achieve in a hypothetical course. Here, too, students must identify experimental and control groups, and independent and dependent variables, which Becker says most students find difficult.

Becker, who has been teaching biology for 24 years, started using case-scenario questions on her biology exams as early as 1989. Completing her pharmacy degree in 1988 played an important role in her thinking about exams. Studying pharmacy, she discovered a wealth of "real-world" situations to use in her classes. Problems students could relate to (such as questions patients might ask about how a drug works) would help them understand the concepts. In one test item, students are told that their patient is taking a drug, A, which is broken down by the liver. He comes in with a prescription for a new drug, B, which is also metabolized by the liver. The question is, "What will happen to drug levels in the blood?" Students must choose the most likely reaction out of four combinations and then indicate which of three explanations best accounts for the reaction.

A typical Becker exam employs multiple-choice and essay questions to give all students an "equal chance" at demonstrating competency despite their different learning styles. Students who find the multiple-choice questions difficult (because of the many distracters Becker uses) know they can compensate on the four essay-format quizzes in the semester-long course. That's "a lot of grading," admits Becker, but she feels what she is doing—adapting tests to make them work better for students—is important.

Like other biology faculty in this compilation, Becker wants her students to master the visual representation of the material. Therefore, she has them draw diagrams on exams. "Students' diagrams show what they don't understand," she says—even after they see slides and the instructor *thinks* they understand, she discovered. Now, even on her multiple-choice exams, students are presented with pictures and asked to identify and to assess, as in the case of embryology, the precise age and stage of development of the embryo shown in the diagram.

Finally, Evelyn Becker uses ungraded and unannounced self-assessment tools to help students do better in biology. Short, concept-oriented questions are intended to help students apply critical-thinking skills to their newly acquired biology understanding. The student is asked to put some principle into words, such as explaining why, as a pharmacy intern, the student affixed a "Do not take with food" label to a vial of "flimflamic" acid, or to write out the steps in glycolysis, to take another example. Although they are not announced in advance, the assessments are not threatening, because they are short, ungraded, open-book, and because the instructor walks around the room giving help.

Although not graded, Becker gives written feedback on the self-assessments, teaching her students what to focus on in their study, and something about presentation of information. On course evaluations, student response to the self-assessments has been overwhelmingly positive.

Becker's colleagues have been willing to try essay and thought-provoking test questions, but those with large classes feel overwhelmed by the challenge of grading so many different exams.

GROUP EXAM "SELF-ANALYSIS"

Lorena V. Blinn, Center for Integrative Studies in General Science, 100 N. Kedzie Hall, Michigan State University, East Lansing, MI 48824; TEL: (517) 432-1560; FAX: (517) 432-2175; E-MAIL: lblinn@msu.edu.

Courses Taught

- Applications of Biomedical Science (an Integrative Studies biological science course)
- Natural Science courses in the biological, physical, and geological sciences

Description of Examination Innovation

Lorena Blinn learned over a decade ago that simply returning graded exams did not do nearly enough to help her general education science students master the introductory chemistry or biology material included in her course.

"My job is to help the student determine why one answer is better than another, not simply to grade 'right' or 'wrong,'" she says. So she shifted the responsibility of diagnosing error to her students. In 1982, she started spending as much as a full 2-hour class period after each exam on what she calls "exam self-analysis." She finds this method particularly well suited to students with skill deficiencies in reading, math, and science, and with low self-confidence.

Blinn's system works as follows:

She returns tests without answer sheets to the class after the exam for "group deliberations," in which groups of three to four students redo the exam together. They may use notes, books, and other sources. Blinn circulates among the groups, listening to arguments without interfering or answering questions. Interestingly, she notices that without their marked answer sheets to bias them, students approach the questions more open-mindedly—more so than when reviewing the exam with their graded answer sheets in front of them.

After these "deliberations," which can take longer than the original exam, Blinn reviews the test in class, calling on groups for answers and explanation. Individual students record the answers. When groups disagree, students debate and resolve their differences. Finally, Blinn answers whatever questions have been raised and returns the graded answer sheets. "I was intrigued, and the students were amazed, by the fact that some of them had argued most convincingly for a particular answer, only to find it was not the one they had put on their answer sheet!"

"Once the system was in place, students stopped complaining about 'tricky' questions," she says. "Instead, they focused on themselves, trying to discover why certain answers—theirs and others'—were incorrect."

Blinn's system allows students to discover for themselves what every instructor already knows—namely, that poor preparation and poor exam-taking strategies account for many student errors. After the postmortem, Blinn finds far more students requesting extra help than when she simply returned the graded tests. It takes at least an hour to do the exam self-analysis in class, but Blinn believes that the science understanding gained, and students' recognition of their problems or deficiencies, saves time in the long run. Because several differently constructed questions (Blinn uses multiple-choice and essay questions that require students to apply what they've learned to new situations) on each of her exams cover the same material, Blinn can discriminate between misconceptions about subject matter and problems of another nature. Early intervention also helps the instructor determine whether a particular student would benefit from the Learning Resources Center, the campus counseling center, or group study with other class members—any of which she will arrange.

In the mid-1980s, several of Blinn's colleagues became interested in her technique and began using it. After reading her *Journal of College Science Teaching* article about this method, other instructors followed suit.[1]

But 10 years later, her institution, like many others, has seen class size increase tenfold. Now teaching assistants control laboratories and study sections, and faculty give more formal lectures. As a result, Blinn has had to abandon "exam self-analysis" during class, though she continues the process in a limited way. She tells students to see her outside of class if they don't receive the grade they expected for their effort and time. During her office hours, the students go over their exams—first on their own and then with her, sometimes in groups of two or three, sometimes individually.

---

[1] "Exam Self-Analysis: Helping Students Accept Responsibility," *Journal of College Science Teaching* (March/April 1985): 424–425.

Today, therefore, it's only the students "who really want to work" whom Blinn sees now. She also doesn't get to hear the class debate questions that helped her understand her students' misconceptions or detect ambiguous exam questions. The group version of this innovation was a public event during which students and instructor "felt we were working on something together." Now students must take the initiative and meet the instructor outside of class.

The loss is significant, the instructor believes, both in terms of students learning the subject and students learning about themselves.

TREE IDENTIFICATION EXERCISE THAT TEACHES USE OF TAXONOMIC KEY

Tee Brower, Professor Emeritus, Department of Biology, Armstrong Atlantic State University, 11935 Abercorn Extension, Savannah, GA 31419; TEL: (912) 927-5314; FAX: (912) 921-2083.

Courses Taught

- Introductory Biology
- Plant Physiology
- Entomology

Description of Examination Innovation

Tee Brower (retired) taught her introductory biology students to use a standard taxonomic key by having them identify native and exotic trees around campus. Students, working first in small groups, became comfortable with the key's figures and glossary by identifying the dozen or so trees Brower has tagged. This required students to put to use the concepts they had learned in lecture. Students also said they liked being outside the classroom once in a while, free to talk with one another about their assignment.

The graded exercise came next. Brower cut branches off three trees the students most likely hadn't seen before and brought them into the lab. Students had to identify the three trees and explain how they made their identification—in other words, what route through the key did they use?

"This exercise trains them to look critically at small details and to think through a problem systematically," says Brower. To reinforce the significance of logical thinking, she gave partial credit, even if the student ultimately failed to make the proper identification, based on how far the student went "correctly" with the key.

"I see this as an exercise," Brower says, "but also as something that students who don't do too well sitting there looking through a microscope can really enjoy and get a lot out of. I'm teaching scientific decision making, and it's not always my A students who are the best decision makers."

With advanced students, the exercise involved making their own keys to a number of unknown plants. "If they've gotten to the point where they can make their own key, they've learned to be critical and look at detail," says Brower. "They also understand the bigger picture of how things are related to one another."

Brower developed this exercise with a colleague 16 years ago and used it ever since. It usually counted as the equivalent of one lab quiz (out of five), sometimes as a source of bonus points on the final. Other instructors at Armstrong use Brower's exercise. It is possible to do similar exercises with any group of living organisms, says Brower.

EXAM REWORKS WITH JUSTIFICATION REQUIRED

Virginia Buckner, Department of Biology, Johnson County Community College, 12345 College Boulevard, Overland Park, KS 66210; TEL: (913) 469-8500 Ext. 3395; FAX: (913) 469-4409; E-MAIL: vbuckner@jcccnet.johnco.cc.ks.us.

Courses Taught

- Principles of Biology

Description of Examination Innovation

Like many others in this collection, Virginia Buckner has allowed students to redo multiple-choice questions they answered incorrectly on an exam in exchange for additional credit. After grading the exams, the multiple-choice questions, along with the Scantron answer sheets, are returned to the students. Students are told how many of their answers were incorrect, but not which ones. They may take until the next class period to rework the questions, using resources such as the textbook, notes, and one another.

Upon observing "knots of students in the corridors" after returning exams for reworking, Buckner became concerned that the method was not providing for careful reconsideration of student answers but, rather, "biology by majority rule." To make students more accountable for their own reworking of the problems, Buckner modified the option and added a new proviso: additional credit would not be given unless students justified in writing the reasons for any changes they made and turned in the exam no later than the next class period. They were still permitted to consult any source they wished and were awarded a half-point—or about one-fourth of the full value of the question—for each reworked correct answer.

Approximately one-third of her 35-member class takes advantage of the rework option. However, the students Buckner most wanted to help through this method—the students who have done particularly poorly—tend not to rework exams. Rather, the option appeals to those students who have missed the least number of problems, such as the high-B student trying to make an A, who returns to class with a near-perfect paper.

Buckner's intention is to encourage her students' efforts to relearn what they didn't understand before, to allow them (and her) to identify persistent misconceptions and misunderstandings of the material, and to get students to recognize how many of their errors are simply careless. She reads all the justifications, and when an answer is still incorrect, she provides written feedback to the student. Their written explanations allow Buckner to feel more confident that students are not simply copying answers from one another. In addition, she doesn't allow students who get near-perfect scores the first time around to rework exams. She holds onto those test papers until all papers are turned in for the second time.

Short-answer items (approximately 30 percent of her exams) are not included in the rework option, partly because of time (she teaches four sections of biology), and partly because students already get to choose which short-answer items to complete. Her multiple-choice questions, assessing recall and comprehension, directly correlate with specific course objectives she includes on handouts for each chapter.

If a student makes more mistakes on the second test than on the first (occurring about 5 percent of the time), the first test score stands.

HANDS-ON ACTIVE LAB PRACTICALS

Sharron A. Clark, Department of Life Sciences, Golden West College, 15744 Golden West College, Huntington Beach, CA 92647-0592; TEL: (714) 892-7711 Exts. 51110 or 52225; FAX: (714) 965-7873;
E-MAIL: Sharron=Clark%MathScience%GWC@banyan.cccd.edu.

Courses Taught

- Introductory Biology
- Introductory Botany
- Global Ecological Studies

Description of Examination Innovation

Laboratory practicals in Sharron Clark's biology and botany classes do not resemble what she calls the "musical chairs" format, in which the professor takes the class from station to station, pointing to slides or pictures, and the students write single-word identifications on their papers. During Clark's practicals, you'll see students walking around the room preparing slides. You'll hear them talking to each other as they gather supplies or clean up their work area. You'll see them walking outside, gathering stems from plants on campus to illustrate phyllotaxy. "They are doing more science than they would in the traditional practical," says Clark, "and the skills they're using during this exam better resemble what the scientist really does than those in the traditional exam."

"Science is a dialogue, not memorizing names of plant parts," says Clark—which is why her practical questions are open-ended and frequently allow for more than one answer. Her students are given a microscope, slides, living organisms (or the "laboratory" of the outdoors), and approximately twelve 3" × 5" cards with multipart questions on them. For example, if the class has been discussing fungi, a student's card may direct him/her to prepare a slide with basidiospores on it. The student would have to know this was a mushroom and be able to find the appropriate part within a mushroom cap. The student would have a choice of using fresh mushrooms or prepared slide materials. Each student gets a unique practical, although the questions are similar in difficulty and content.

"To believe that students who can translate tasks into memorized words are the ones who have really learned is to neglect the richness of hands-on experiences in the laboratory," says Clark, who also spends much less time setting up practicals (she estimates two–three hours before) and much more working with students than she did before she first started using this method. She meets with each of her 30 students *individually* to grade their practicum. Each student has a scorecard with a list of all parts of the 12 questions. Clark writes down the students' scores for each answer as they move together through the stack of cards. "I learn more about how the students' minds work while we discuss answers, and the students get instant feedback," she says.

In the past, 25–50 one-word identification tests (e.g., using fetal pigs with numbered flags stuck into specific veins, muscles, and nerves) took three–four hours for Clark to set up and about 1 hour to administer. Now, her setup time has been reduced to less than 1 hour (as students do their own), and it takes between 1 and 2 hours for Clark to test all her students (with increased sensitivity to a range of learning styles). Students also generate test questions: During laboratories, Clark has students jot down questions (and she does the same) to be used for discussion after the lab period. Clark selects the best, most representative questions for the exams.

Clark's students find it less stressful to be moving around the room doing work than standing in one place for an hour taking a grueling identification test. Of course, now they must interact one-on-one with the professor during the exam, which opens many of the students to a new level of exposure in the science classroom. This may be one of the more powerful benefits of Clark's innovation.

Her colleagues, though interested in the method, have not adopted or tried it in their classrooms.

COOPERATIVE STUDENT GENERATION OF ESSAY-EXAM QUESTIONS

Georgianna Glose, Community Service Center, New York City Technical College, 300 Jay Street, N-422, Brooklyn, New York 11201; TEL: (718) 260-5117; FAX: (718) 237-0905 (call before sending).

Courses Taught

- Introduction to Gerontology
- Community Organizing and Development
- Introduction to Human Services

Description of Examination Innovation

Georgianna Glose has her introductory gerontology students compose essay-exam questions to help them develop their thinking and writing skills. After each unit, students in already-formed study groups share the two essay questions they have written with the others in the class. The instructor selects six student questions (substituting her own question if a significant content area has been omitted) for a unit review done in study groups before the examination. The individual in-class exam is made up of four of these six questions—two selected by Glose and two selected by the students.

"By letting them compose questions for the test, students gain a sense of control and responsibility for their learning," says Glose, who observes their active development of responses to questions in groups, including bringing collateral resources into class to assist them in their question preparation and responses. The method also provides students for whom English is a second language with a less intimidating testing situation.

Through the use of reaction papers in class, Glose has concluded that students come to class more prepared to work with each other. "It's important to *prepare* students for all-essay exams, however," she says. "I give them sample essay questions and examples of poor as well as good student responses, all of which we review at the start of the term."

Glose says her students show more than a rote understanding of the material and are able to apply theoretical ideas to practical situations. At first, her students are "very anxious" about the all-essay exams, but once they begin to understand how to prepare their notes and answer the questions—assured, too, that there will be "no surprises"—students concentrate on learning the content.

OPEN-BOOK EXAM REWORKS

Eileen Gregory, Department of Biology, Rollins College, 1000 Holt Avenue, Winter Park, FL 32789-4499; TEL: (407) 646-2430; FAX: (407) 646-2479; E-MAIL: egregory@rollins.edu.

Courses Taught

- Microbiology
- Biology for Teachers
- General Biology

Description of Examination Innovation

For four years, Eileen Gregory has been using a modified version of an examination grading practice developed by Eleanor Siebert at Mount St. Mary's College. The first time Gregory returns students' exams, there are no marks on the multiple-choice questions—just a number indicating how many of their responses in the section were incorrect. Students have 24 hours to redo any multiple-choice questions they wish and are encouraged to use their text, notes, and each other as resources. When they turn in their tests the second time, Gregory adds points to their original exam score, the amount depending on how many answers are now correct and on the value assigned to each question, which varies from test to test.

Gregory says her students have responded quite favorably to this technique.

"They say that going over the material again helps them to understand it in more depth than before the exam," she says. And because most of her multiple-choice questions involve analysis and critical thinking, the exercise gives them practice in these important skills. It also encourages cooperative learning and makes assessment an integral part of the learning process.

"Students who did poorly on the multiple-choice section of the test benefit more from this activity than those who did well," says Gregory, "but both the students and I feel this is a better method than adding points to everyone's score in a curve."

Gregory recently altered this exercise when other faculty in her department began using the technique in their lower-level classes and found that many students were reworking their exams in large groups. Papers were coming back with correct answers but too little written work on them. It was hard for faculty to be sure each student could do the problems individually, so Gregory began insisting that students work only in groups of two or three on their exams. Each student puts his/her collaborators' names on the exam, and all are held to an honor code that they did not receive help from other students. In this way, individual accountability is retained.

MICROTHEME WRITING ASSIGNMENT

Ceil Herman, Biology Department, New Mexico State University, Las Cruces, NM 88003; TEL: (505) 646-1325; FAX: (505) 646-5665.

Courses Taught

- Principles of Biology
- Endocrinology
- Human Physiology
- Histology

Description of Examination Innovation

Two years ago, Ceil Herman starting using "microthemes," or short writing assignments, to assess students' understanding of the main points of her lectures for Principles of Biology and Human Physiology. Faculty in the biology department at New Mexico State had reported good results using methods they'd learned in a "writing across the curriculum" workshop for faculty. Herman now uses writing, in-class and on exams, as a testing and teaching mechanism. She can be considered a convert to the principle of extended writing in the science classroom.

Her students purchase a set of 3" × 5" notecards, which they come to use extensively for microtheme assignments. In the last 5–10 minutes of lecture (although sometimes in the middle), Herman poses a question to the class, and students answer it on their cards, usually in a couple sentences. Most often, Herman asks students what the main points of the lecture were; occasionally, she'll give them a science article in class, and the next day, they have to answer a few questions about its main points.

Students who want their cards reviewed by the professor put their answers on colored notecards. At the beginning of the term, 50 percent chose this option. The others put their answers on white cards, receive credit for their microtheme, and do not get them returned. By the end of the term, 75 percent choose to have their cards returned. Not only do students respond to Herman's questions, but they offer feedback about her lectures and ask specific questions about those parts of it that confused them.

The instructor gives written feedback on all the colored cards and goes over the answer to the question in class when she returns the colored cards. If many students have the same concern or misunderstanding, Herman gives the issue more time. "The method is flexible enough to let me do what I want," she says.

In the Principles of Biology class, with between 150 and 200 (primarily) freshmen, Herman gives credit for the microthemes but does not grade them. However, she does insist they write their answers in essay style, not in list form.

In her human physiology course, with approximately 80 sophomores and juniors, Herman's microtheme questions focus more extensively on science content. Typically, a question will have an opinion part, such as how students feel about a daughter killed in a car accident whose heart is transplanted to her father to save his life. A second part will ask students about the science of the same issue, such as the technical problems associated with emergency organ transplants. The first part is worth a certain number of points but isn't graded, whereas the second part is graded for accuracy.

Students who write microthemes tend to do better on essay exams than those who don't, reports Herman. In her microthemes section, the class average on a short answer/essay exam was 73 percent, whereas in her other section, which doesn't do microthemes and takes only multiple-choice exams, the average was 63 percent. Although the difference is due to more than the use of microthemes, Herman says her students tell her they *feel* they're learning more biology through their writing.

Herman has also changed her exams to include more essay questions than ever before. Although these take longer to grade, multiple-choice exams take longer to prepare.

What convinced Herman to move toward essay exams was a dyslexic student in her class who said she'd gotten a B in organic chemistry by just circling the same letter on every multiple-choice test. "She knew almost no organic chemistry, of course, because she could not see the structures correctly," said Herman, who wanted to teach her students more. Her first exams were more general than the ones she writes now. Now, her questions are more focused and open-ended. For example, in 1992, she asked her students to explain "what would happen to the environment if there were no or fewer green plants than there are?" Now, an 11-part question about photosynthesis will require that students locate the process in the plant and distinguish the steps involved in plant and cyclic and noncyclic photophosphorylation.

One of her colleagues occasionally uses microthemes in her class with approximately 300 students, but not all the time. Herman, too, would use this method if she taught large classes. The "writing across the curriculum" workshops given every year at New Mexico State encourage faculty from all fields to use writing more often. "Beyond everything else, having them write and read their writing is a great way of maintaining contact with students," says Herman.

OPEN-BOOK OPTION AT A POINT "COST," MODIFIED CURVE

William N. Hudspeth, Department of Biology, Northwest Missouri Community College, Senatobia, MS 38668; TEL: (601) 562-3379; FAX: (601) 562-3911.

Courses Taught

- Introductory Biology
- Introductory Botany

Description of Examination Innovation

William Hudspeth sometimes gives open-book/note exams, but it *costs* his student one point per minute of "open time." For 15 minutes, students work on their exams without notes. Then, they are permitted to open their books. Those who choose to do so raise their hands and the instructor clocks them in by noting the time when they raise their hands and again when they close their books. The technique gets students to appreciate the importance of exam preparation.

"My best students need only 1–3 minutes of 'open time' because they know where to look in their text and notes. Poorly prepared students will literally do themselves in," says Hudspeth. Unfortunately, he finds it nearly impossible to administer this exam option to groups larger than 20 students, his usual class size.

Hudspeth also uses what he calls a grading "bump" to improve student morale when scores are low. By taking the square root of a student's raw score and multiplying it by 10, he can still get a fair distribution of grades. He says when raw scores are low, this system benefits most the student scoring in the middle range.

Another curve variation he uses is doubling the first digit and adding that to the raw score: a 62 becomes a 74. With this method, high-scoring students get the most help, very low-scoring students get the least. Both techniques are designed to scatter clumped low grades in the passing range. The raw score is the student's means of placing him/herself in the class distribution. From there, says Hudspeth, "we work it to some letter grade."

POSTEXAM OPPORTUNITY TO DEBATE A GRADE

Lloyd A. Jones, Department of Biology, University of Toledo, Toledo, OH 43606; TEL: (419) 530-4596; FAX: (419) 530-7737;
E-MAIL: LJONES@utnet.utoledo.edu.

Courses Taught

- Introductory Biology
- Honors Biology

- Field Botany
- Domesticated Plants
- Plant Physiology

Description of Examination Innovation

After handing back his short-answer and essay exams, Lloyd Jones encourages students who disagree with his grading to come to his office to explain how he "misinterpreted" their answers. Jones warns them to study the material related to the question before making this attempt. If the complainant can demonstrate competency in the subject related to the question, Jones says he's "easy to convince" that he graded too harshly, and he awards extra points. But if they can't answer Jones's questions about the material, no points are awarded. Jones's goal is to teach his students, some of whom do not take any other essay exam during their college career, how to articulate basic biological concepts.

"I feel my role is to encourage, not block their efforts," Jones explains. But the option can become overwhelming for the instructor when students avail themselves of it. In a class of 250–300 students, 15–30 for each exam, or 100 in total, will come to see him over the course of the quarter, often for half an hour at a time. More start coming in toward the end of the term, and they're "more desperate," says Jones. In cases in which the student isn't prepared, Jones says the meeting becomes a "reality check" for the student. Some students, he admits, come only to impress him, but most have revisited the material.

"Is my method fair?" Jones asks himself. "To be sure, the motivated students get the most out of the opportunity. But my goal of having students really learn the basic biology also comes closer to being met."

TAKE-HOME SECOND-CHANCE OPTION FOR HALF CREDIT

Roger M. Knutson, Department of Biology, Luther College, Decorah, IA 52101; TEL: (319) 387-1117.

Courses Taught

- General Botany
- General Ecology
- Plant Taxonomy

Description of Examination Innovation

In the past, Roger Knutson's exam questions in general botany and general ecology were so difficult that the mean grade on his tests was about 50 percent.

Although he curved grades, his students were troubled by their low absolute scores. Knutson, too, had to admit that not much learning went on if a student couldn't answer a question and learned nothing new in the process.

So, to get students to really learn what he considers the most important material in the course, Knutson now gives students a second chance. He returns their graded exams, indicating which questions were answered incorrectly, and invites students to answer any missed questions at home for an additional half credit. (When partial credit has been awarded, a student receives half of the remaining points missed.)

For students, an added incentive for the take-home second attempt, besides the chance of raising their grade, is that the instructor tells them some questions from earlier exams—usually a third—will comprise a proportion of the final. "The only people who don't do the take-home retake are those who already answered 95 percent correct," says Knutson. Even though he grades take-home differently, insisting that students write deeper and better organized answers than if they had answered the question in class, students usually get the problems right with the extra time and resources.

As a rule, all of Knutson's questions ask for brief explanations or the experimental basis for major ideas. "In a sense they're nearly objective questions because I'm usually looking for three to four pieces of information," he says.

In course evaluations, students are generally positive about the second-chance option. They acknowledge the extra time it takes to answer the kinds of questions Knutson poses, but since they are biology majors, they value the deeper and more comprehensive understanding that they gain.

WEEKLY QUIZZES, LABELING DIAGRAM QUESTIONS, PREQUIZZES

Jane Kolunie, Department of Biological and Physical Sciences, Barton College, College Station, Wilson, NC 27893; TEL: (919) 399-6470; FAX: (919) 237-4957; E-MAIL: jkolunie@barton.edu.

Courses Taught

- Principles of Biology (and lab)
- Structure and Function of Man (and lab)
- Animal Behavior
- Physiological Psychology
- Human Sexuality
- Neuropsychopharmacology

Description of Examination Innovation

Two years ago, Jane Kolunie asked her anatomy lab students what they thought might help them do better on her practicals, and they told her: weekly quizzes.

In a course comprised of more than 75 percent vocabulary and visual information, frequent testing can keep students on track. To that end, Kolunie requires that her students do some drawing every week, and then she gives a 10-minute quiz the following week, in which a section of the body is pictured, with numbered parts (such as the axial skeleton or the skull) that the students must correctly label. Test diagrams are based on the ones students labeled and reviewed the week before in their course workbooks. Kolunie randomly calls out approximately five specific structures in the diagram that the students must label.

Although her colleagues think Kolunie is "nuts" because of the additional work she has assigned herself, she persists with the quizzes and the complicated "relay-style" practicals that other professors streamline into multiple-choice exams. In Kolunie's practicals, students move through 20–25 different work stations, answering questions based on models, dissections, slides, and diagrams. Kolunie insists on using realistic visual aides because she is convinced that most students do not master anatomy by means of traditional paper-and-pencil exams alone.

"How can you learn whether a student can distinguish biceps and triceps with a multiple-choice question?" she asks. Students grasp anatomical concepts better, according to Kolunie, if they examine a diagram or model (or subject) and make the distinction than if she asks them, "Which muscle moves the shoulder up?" essentially a vocabulary question.

"It would be easier for me to use nothing but photocopied diagrams," she adds, "but students learn better, and the material stays with them longer, if they're tested in a hands-on style."

It was one of her student's inability to draw a muscle after a detailed course lecture on muscle structure 3 years ago that convinced Kolunie she needed to teach her students to *see* anatomy, rather than merely to memorize terms and associated functions. This is especially important for her nursing students, who will need to know how to *recognize* these structures. The weekly, visually oriented quizzes comprise between 25 and 30 percent of course grades, and although students occasionally complain about having so many, almost all of them say the quizzes are helping them learn the material. Kolunie has even been asked by students to make them worth more points!

Kolunie has adapted another student suggestion in her general biology class (a required course for science majors and a core requirement for virtually all others at Barton). To combat students' lack of preparation for laboratory, she gives a "prequiz" every class (when an exam or quiz is not already scheduled) based on a general question from either the reading in the textbook or in a previously distributed laboratory handout. The 10-minute "prequiz" helps ensure that students have read and understand the overall concepts in the lab by the time they get to class. Kolunie's laboratory exams include a practical section, in which students examine and identify specimens, worth 25 percent.

Kolunie is obtaining good results with these techniques, and her attendance rates have gone from in the 80s to 100 percent. Few students are tardy, knowing a quiz will be given. Most important, her students tell her they are understanding more.

MODIFIED MASTERY SYSTEM USING PERFORMANCE OBJECTIVES

J. Richard Kormelink, Biology Department, F-202, Delta College, University Center, MI 48710; TEL: (517) 686-9262; FAX: (517) 686-7320;
E-MAIL: rkormeli@cris.com.

Courses Taught

- Principles of Biology

Description of Examination Innovation

Over the past 20 years, Richard Kormelink has developed and improved a system for testing performance objectives in his biology course. Each objective is tested by means of three multiple-choice questions on a particular exam. On the first day of class, Kormelink hands out a list of the 120 objectives, on lecture-based, theoretical, or textbook-oriented information. Success at meeting the objectives comprises one-half of the students' grades. The other half is based on laboratory activities. Best of all, students no longer complain that they did not know what to study.

Examples of objectives include the following:

Objective 3: "Recognize a definition for scientific method, observation, hypothesis, empirical evidence, experiment, variable, theory, control group, and experimental group, and how these terms apply to the process of science."

Objective 7: "Determine the mass number and/or atomic number of an atom and the number and locations of particles within the atom when given the information on a periodic table of the elements."

The instructor gives four exams during the semester on which students must correctly answer two out of the three questions to meet each particular objective. Thirty objectives (tested in the multiple-choice format for ease of grading, with a total of 90 multiple-choice questions) are on each of these exams.

The scoring sheet allows the instructor to scan groups of questions by objectives so that he can accurately measure achievement by objective. Since the laboratory portion of the class allows students to write, assimilate, and integrate information, the instructor does not feel that his 100 students are in any way shortchanged by the use of multiple-choice questions on exams.

After the first two exams, and then after the last two, students are given an optional retest, with three new questions for all 60 objectives. On the answer sheet, the students indicate which objectives they're "retrying." In the retest, the student must answer all three questions correctly to meet the objective.

In the past, students tended to take the retest without much additional preparation. To counter this, Kormelink recently changed the way in which students qualify for retests. He provides a "ticket," called a "biology bonus buck," to a student who has done particular activities to better prepare him/herself for the test or retest. These activities include (but are not limited to) 30 minutes in the multimedia learning lab practicing multiple-choice questions; viewing videotapes of relevant materials, spending time with a PLATO system, performing additional laboratory exercises, spending time reviewing previous tests in the instructor's office, and working in a tutoring session.

Each of these activities is equated to a "biology buck" that can be accumulated and used to "pay" for the opportunity to retake the test questions on a particular objective—one "buck" for each objective. This change has reduced significantly the number of students who avail themselves of the retest opportunity, and the quality of preparation has in turn gone up.

ESSAY QUESTIONS ON CURRENT EVENTS, ADVERTISEMENTS, AND POPULAR SCIENCE ARTICLES

Jay B. Labov, Department of Biology, Colby College, Waterville, ME 04901; TEL: (207) 872-3329; FAX: (207) 872-3731; E-MAIL: jblabov@colby.edu.

Courses Taught

- Introduction to Biology
- Mammalian Anatomy and Physiology
- Animal Behavior
- Topics in Neurobiology

Description of Examination Innovation

Jay Labov uses short essay questions on his examinations, even in an introductory biology course, with nearly 200 students. Students write about current biology issues in the news, analyzing scientific information from advertisements and popular science articles. For example, after discussing the physiology and endocrinology of human reproduction in class, Labov has included on an exam an ad taken directly from *Time* magazine that describes a product that is said to predict when a woman is most likely to ovulate. Labov asks his students to use their knowledge of reproductive physiology to hypothesize how the kit works. They must then defend their answer.

In Topics in Neurobiology, an upper-division course, after discussing the physiological basis for nerve impulses and the membrane structures and physiology involved in generating these impulses, students are given a hypothetical experiment to conduct. The new drug they are investigating changes certain characteristics of a nerve cell's ability to generate impulses. Results from control and experimental nerve cells in four experiments are juxtaposed on the page. Based on this information, students must develop a hypothesis about how the drug is influencing the nerve cell and use the experimental evidence presented in the question to support their hypothesis.

In several classes, Labov also employs questions that present two opposing views on a scientific debate. Students are asked to design an experiment that would resolve the debate and explain how their experiment would do so. For example, in his Animal Behavior class, Labov presents different theories to explain feeding behavior of young mammals—one from psychologists, the other from ethologists. The students must design an experiment that would distinguish between the competing theories and then defend their answer.

Students find Labov's exams to be "different" than most they have taken in other courses. Aware that their unfamiliarity with his methods could make the exams unduly challenging, Labov always puts copies of the previous year's exams on reserve at the start of the term in the campus library or, more recently, in an electronic newsgroup established for the course, so students can become accustomed to his style of questioning.

Labov's work contradicts the view that an instructor cannot cover on in-class examinations material that would be encountered in a laboratory, or cannot ask the kinds of questions that require creativity and a healthy appreciation of both the powers and the limitations of the scientific method.

CUMULATIVE, SELF-AUTHORED MULTIPLE-CHOICE FINAL AND TAKE-HOME CONCEPTUAL EXAMS

Randy Landgren, Department of Biology, Middlebury College, Middlebury, VT 05753; TEL: (802) 443-5680; FAX: (802) 443-2072;
E-MAIL: landgren_randy@msmail.middlebury.edu.

Courses Taught

- Introduction to Organismic Biology

- Genetics and Evolution
- Plant Physiology
- Plant Anatomy
- Introduction to Plant Sciences
- Writing on Nature
- "Good, Right, and True" (science ethics)

Description of Examination Innovation

Randy Landgren has created two new kinds of examination for his students in a first-year college course in genetics and evolution at Middlebury College. One is a "cumulative, self-authored multiple-choice final examination"; the other is what he calls a "progressive pyraMIDD examination" consisting of two to four highly conceptual questions. The course itself uses many of the strategies of Eric Mazur's Harvard physics course, with class lectures punctuated by periods of review and reflection, and collaboration with peers.[2] Landgren's class of 80 works in five laboratory sections of 12–20 students each. Sections break down into smaller study work groups of 3–5 students.

Landgren has always designed his graded exercises in the course (essays, midyear examinations, and laboratories) to encourage review, to reinforce students' grasp of conceptual material, to reward both empirical comprehension and creativity, and to help students teach one another. The two exam innovations, which take place during the scheduled 3-hour examination period, are intended to extend these goals.

The essence of the self-authored multiple-choice final exam is explained to students 2 weeks prior to the examination itself. Each student composes four multiple-choice questions of his/her own design, consisting of a stem, a correct answer, and three distracters. The questions are in four different areas of the course subject matter. The questions are then presented to small or large groups for "editing." The editing process is part of Landgren's teaching strategy, and he begins by modeling a few edits himself. First, the proposed questions are sorted by subject matter and then critically examined along the following criteria:

1. What is the biology being tested?
2. What answer do the "editors" believe to be correct?
3. Why is each of the distracters incorrect?
4. Is the language/grammar/tone consistent?
5. Is there room for misinterpretation based on cultural, racial, experiential, or gender biases?

[2] See "Students Teaching Students: Harvard Revisited," in Sheila Tobias, *Revitalizing Undergraduate Science* (Tucson, AZ: Research Corporation, 1992), pp. 114–122.

Finally, the edited questions are posted in a database, and paper copies are placed in several locations. Landgren constructs the final examination out of 60 of the "best" questions. In addition to answering these questions, students have to identify, during the final examination, five areas of the curriculum that were not tested on the exam. Landgren considers their ability to articulate these specific areas to be one measure of their overall understanding of the course material.

Initial student reaction, reports Landgren, was skeptical. The process was perceived at first to be trivial. Students thought that both the writing of the questions and the taking of the examination would be easy, but as they began the writing and editing process, they quickly recognized the importance of identifying misconceptions and began to strengthen their grasp of the empirical material. Half the class ultimately chose not to come to the last editing sessions and to accept their "authoring" grades as their "editing" grades (by skipping one or more of the last three editing sessions). But those students performed less well both on the examination and on the task of identifying areas of the curriculum not adequately covered by the examination.

The second innovation, the "progressive pyraMIDD examination," is also explained in advance to students, and they are given take-home conceptual problems to prepare them for the kinds of questions they will be given in class. The final is divided into four time segments: During the first 30 minutes students are in a closed-book mode, working individually, and are encouraged to speculate and sketch out ideas rather than to give complete answers. Grading of this section is liberal, and partial credit is granted generously. During the second 30 minutes, students work with open books, again by themselves. During this portion of the examination, Landgren wants his students to find support for their ideas in the written materials they have brought in with them. Partial credit in this section is less generous, and students are rewarded both for realizing any weaknesses of their initial ideas and for documenting the strengths of those ideas.

During the third time segment (60 minutes in length), students work in groups of three to five students, and each group is expected to produce a single set of polished answers. Landgren is "stingy" with partial credit and expects to see a good grasp of empirical and concrete aspects of the material in the responses. And in the last 60 minutes, the full group discusses the small-group questions just completed. This (ungraded) time period is meant to air flaws in the conceptual and empirical parts of the answers given by the smaller groups. (The products of each of the time segments are graded separately.)

Initially, students felt this examination was too long and cumbersome, but once they became familiar with the instructions and saw that they were graded fairly, students came to appreciate that a wider spectrum of abilities was being evaluated.

The advantage to both systems, according to the instructor, is that higher-order thinking and collaborative effort is rewarded. The disadvantage is that grading is difficult: There is more to grade, and the grading schemes are complex. One other faculty member at Middlebury has adopted both methods in a course team-taught with Landgren. Others have expressed interest in the techniques but balk at the logistics and the subjectivity of aspects of the grading.

STUDENT-CORRECTED TRUE–FALSE QUESTIONS, "JUSTIFICATION" AND "RELATIONSHIP" QUESTIONS, STUDY QUESTIONS

Debbie Meuler, Department of Biology, Cardinal Stritch College, 6801 North Yates Road, Milwaukee, WI 53217; TEL: (414) 352-5400 Ext. 329; FAX: (414) 352-7516; E-MAIL: dmeuler@acs.stritch.edu.

Courses Taught

- General Biology (for majors)
- Principles of Biological Sciences (for nonmajors)
- Immunology
- Cell Physiology
- Developmental Biology
- General Physiology

Description of Examination Innovation

In Debbie Meuler's lower-level course examinations, traditional formats, such as true–false, multiple-choice, and short-answer questions, have been modified to improve their evaluative efficacy.

For true–false questions, her students must correct false answers so that they read correctly. This obviates the 50–50 chance of getting true–false questions right just by guessing. Students do not get credit for a correctly identified false answer unless they change the statement into a correct one.

Meuler also uses what she calls "justification" questions, a variation of multiple choice also used by others in this collection. In these questions, students must not only select the correct answer (from four or five selections) for each question, but to get credit for the question they must justify why each distracter was accepted or rejected. This forces students to articulate their thinking processes. Beginning students many times can pick out the right answer but have difficulty explaining why that answer was right and the others were wrong. This type of question also helps student and instructor determine how well the student understands the material.

The following is a typical justification problem:

Normally, cells from plant A have 24 chromosomes in every cell. You observe one cell from Plant A and it has two sets of 12 chromatids. This cell is most likely in what stage of cell division?

a. Metaphase of mitosis
b. Anaphase of mitosis
c. Anaphase I of meiosis
d. Anaphase II of meiosis
e. Interphase

The questions are usually worth 8 points (if there are 4 distracters). Students get partial credit for attempting to justify their responses to each distracter. From such justification problems, both teacher and student get more thorough and complete feedback than from standard multiple-choice test items.

Another type of question Meuler frequently uses on exams is the "relationship question." Instead of asking a student to define some terms, she asks the student to tell her how the terms are related. For example, she will ask, "What is the relationship between enzyme, substrate, and active site?" If a student merely defines the terms, the answer is marked incorrect. This type of question helps Meuler see if a student understands the whole picture and hasn't just memorized terms.

For final exams, Meuler distributes study questions that are global in nature. For example, questions asks students to explain why cell theory is a unifying concept of biology, or to describe what happens to a steak sandwich as it passes through the digestive tract. These study questions help students "narrow down the mountains of information" Meuler gives them, and helps them focus on what she thinks is important and ought to be remembered 6 months after the class ends. Questions for her final exams are based solely on those study questions, benefiting students if they review the study questions before the exam. Meuler is still surprised that many students do not do so.

In upper-level courses (with small enrollments), Meuler's final examinations are take-homes that ask students to put together a large body of information. The exams usually include four or five essay questions that could not be completed in the allotted classroom time for finals. A typical example requires students to describe the development of an amphibian from fertilization through neurulation. Given that each completed exam is 15–20 pages, it would not be possible to use this exam in larger classes.

Meuler has been using these innovations for over 5 years. Students find the justification questions challenging, which is why Meuler introduces this question format on a weekly quiz given before the first test. She reviews acceptable justifications carefully. Because this question takes so long to grade, Meuler uses only one per exam. Other faculty at Cardinal Stritch College use the true–false and relationship questions, but not the justification question because it is so labor intensive.

TWO EXAMS FOR ONE GRADE

David Netzly, Biology Department, Hope College, Holland, MI 49423; TEL: (616) 395-7718; FAX: (616) 395-7125; E-MAIL: netzly@hope.cit.hope.edu.

Courses Taught

- Plant Morphology
- Plant Physiology
- Plant Pathology
- General Biology

Description of Examination Innovation

Borrowing an idea described in a critical-thinking seminar, Netzly has found a way nearly to double student preparation for exams. The procedure is quite simple: Netzly writes two "equally difficult" tests, with different questions on the same material. Students may take one or both exams. If they take both, the better score is recorded. Unfortunately, the method also doubles grading time. With a class size for the sophomore/junior level biology course that has doubled in the past 2 years, Netzly has had to forego his new idea for the present. He will use it again when he figures out a way to do the grading.

Netzly grades the first exam a day or two after students take it, then has each student come into his office to pick it up. Together, instructor and student discuss the score, including Netzly's written comments and the student's areas of difficulty. Then the student takes the first exam home to use as a study guide if he/she chooses to take the second test.

Do students study more, we asked Netzly, or just take the second exam hoping they'll get lucky and do better? Students have told him they study the material a second time for the retest, using the first as a guide, especially to the abstract concepts they didn't master before.

In this way, the students and Netzly use testing as part of the instructor's teaching function. Although he works hard to make the science clear to his students, sometimes, says Netzly, "it just doesn't click." These exams provide a training experience. It is as if someone is watching over the student's shoulder, and the student learns by repetition.

Students say they like the system because they're learning more by having to study twice, and they very much like having a backup. Netzly says the method is especially helpful to students who "clam up" under pressure. They relax more knowing they have a second chance.

Other faculty are intrigued with the double-testing idea, but Netzly doesn't believe any of his colleagues have adopted it, presumably because of the workload. If his class size is reduced, Netzly says he'll reinstate his innovative testing policy.

ESSAY QUESTIONS, GROUP ORAL EXAMS FOR TEACHER PREPARATION

Valerie Keeling Olness, Department of Biology, Augustana College, Sioux Falls, SD 57197; TEL: (605) 336-4720; FAX: (605) 336-4718;
E-MAIL: olness@inst.augie.edu.

Courses Taught

- Biology and Human Concerns
- Science Methods Courses (elementary and secondary)

Description of Examination Innovation

For the past 3 years, Valerie Olness has been doing a great deal of cooperative work with preservice elementary school teachers, and exams are no exception. In an effort to model alternative and authentic assessment, she has used all types of exams, from multiple choice to essay to oral exams. Her students say they *learn* from her exams and "no longer dread" tests.

Every semester, her students are given two individual exams in addition to four group exams. The first group exam is a short-answer, multiple-choice, and open-ended essay exam, for which Olness breaks the class's cooperative working groups of three into pairs and has each pair produce an answer. Then pairs are put back into groups of three, and the question is answered again, now with more students' perspectives shared and incorporated. Theirs is the final product turned in for a group grade. By this method, each student does the same problem twice, with different input from different fellow students—a meaningful learning experience.

The second group exam is based on a "concept map," a diagramming method to show relationships between concepts and ideas. The class period before the exam, students are given a list of 50 concepts that have previously been discussed, and each student is asked to sketch out a brief concept map to bring to the group exam. In the exam, each student presents his/her concept map, and a final group concept map is produced. When students turn in their final product, they also turn in each individual concept map—"a means of checking for freeloaders," says Olness. These "maps" are time-consuming to read but are quite revealing: misconceptions in relationships are readily apparent. No letter grade is given, but misconceptions are dealt with, through alternative assignments, until the instructor is satisfied that some "conceptual change" has occurred. Despite knowing that no letter grade will result, the students do not take these exams lightly. "Perhaps the prospect of 'eternal assignments' is enough of a deterrent," says Olness. Comments such as "Wow, did you ever make us use our knowledge!" are a testament to the value of this type of assessment.

The third group exam is a group presentation on a particular topic using a double "jigsaw technique." Each group is given a different topic and, within groups, each member is responsible for a particular set of material. Olness sets expected criteria but leaves room for creativity in both presentation style and "other interesting things." The final grade for this exam is a composite of the group grade and individual effort that, so far, has eliminated individual student "freeloading." Following each group presentation, the instructor and other class members ask clarification questions, which generally ensures student preparation. Further accountability presents itself in the individual exams. Thus, students are responsible for both their own and others' learning.

Finally, Olness gives a final group oral exam based on four or five broad "fairly value-laden" questions. Each individual group member must prepare answers to *all* questions. During the final, which takes about 1 hour per group, students draw numbers to see which question each will address. After the question has been addressed, the other group members rebut and the original student responds.

"These turn out to be rather interesting, informative discussions, and quite valuable for the students," says Olness. As she surveys the groups completing the exams, she notices heated discussion and students taking turns writing answers. She feels their final answers represent a group consensus following a thorough group discussion, a kind of "compromise," instead of a narrow or impartial answer. However, if any members of the group do not agree with the prevailing opinion, they may dissent, in writing, by justifying their reasoning. If one student is dominating the group discussion or doing all the writing, Olness will step in. "By not talking or writing, students are not thinking," she says. During the exam, Olness also identifies unprepared students and discusses how they could modify their study habits.

Olness says almost all of her students are excited about the group exams, both because the tests reduce anxiety, and because they empower students by forcing them to articulate the science they've learned. "Group exams give them a better handle on the concepts," she says, "and the students say they're learning more."

In grading essay and short-answer questions, Olness reads everything her students write, paying particular attention to evidence of misconceptions. She either writes detailed clarifications or (for gross misconceptions) assigns alternative assignments until there is evidence of some reconceptualization. "It's a lot more work," she admits, and probably wouldn't be possible if her class size were not limited to 27. For the group oral exams, she uses a criteria sheet that the students have received to guide them in preparing for the exam.

Because of the additional work and class size, other faculty in Olness's department are hesitant to use these experimental methods with biology majors. They also feel strongly about individual student accountability. They acknowledge that group exams can be useful, but only with education students, not biology majors. However, Olness's experience with preservice elementary teachers is leading her to employ cooperative exams in her Biology 110 class, a nonmajors class, in which she already uses extensive journal writing and cooperative learning.

OPEN-BOOK, OPEN-ENDED ESSAY EXAMS

Sandra Palmer, Department of Natural and Social Sciences, Cazenovia College, Cazenovia, NY 13035; TEL: (315) 655-9446 Ext. 146; FAX: (315) 655-2190; E-MAIL: sandy503@aol.com.

Courses Taught

- General Biology
- Genetics
- Aquatic Biology
- Scientific and Technological Literacy
- Death and Dying
- Human Sexuality
- Energy and Environment
- Science and Public Policy
- Microbiology

Description of Examination Innovation

For nearly a decade, Sandra Palmer has given open-book, open-ended exams in all her biology classes. Students can bring any materials they want to the exam. Questions are in either essay or short-answer format and require students to synthesize and apply what they've learned in creative ways. Therefore, there are no "answers" for students to look up in a book, nor is there necessarily one correct answer.

For example, one exam question asks students what they think would happen if (1) DNA duplicated without error all the time; (2) organisms reproduced without meiosis; (3) all the DNA you have is in one pair of chromosomes instead of 23 pairs; and (4) you had no DNA repair mechanisms. Another asks students to explain what their textbook author meant when he/she wrote, "Either we make a global effort to limit population growth in accordance with environmental carrying capacity, or we wait until the environment does it for us." Palmer's goal with such questions is to show students that learning biology isn't simply memorizing facts.

Palmer stopped using multiple-choice questions on her exams because she wasn't getting at the deeper level of thinking she wanted from her students. Her new method requires a lot of grading time, but she is willing and able to do it, even with 40 students in her general biology course. Grades are based on a point system that is absolute. Palmer says that "a curve sets up an unhealthy competition within a class." If everyone deserves an A in her class, they get one. To further reduce student anxiety, students are given "all the time they need" to finish exams, though few need more than a class period. Students do not complain about the writing required for her exams. Some science faculty use short-essay test questions, but not open book. Palmer emphasizes that faculty who teach smaller classes—even as "small" as 40—can be more creative in testing and assessment than those who teach large lecture courses.

GROUP EXAMS

Paul A. Rab, Biology Department, Sinclair Community College, Dayton, OH 45402; TEL: (513) 226-2823; FAX: (513) 449-5192; E-MAIL: prab@sinclair.edu.

Courses Taught

- General Biology
- Introduction to Genetics

Description of Examination Innovation

Paul Rab's group exams, given to honors general biology students, are closed-book, 50-minute, multiple-choice tests, just like his individual exams. Groups of three do the exam, turn it in, and then Rab returns them the following class period with a composite score but no marks on individual questions. The groups discuss the exam for 10 minutes and change any answers they wish. The exams are then collected and given a final grade—the only one that counts—by the instructor.

Rab purposely limits group discussion to 10 minutes—not enough time for groups to radically change their answers—because his goal for the exercise is to get them to work together. Changed answers must be accompanied by a rationale. He hopes that students will have looked up the material they missed the first time around before the next class period.

Group questions, even though they're multiple choice, involve application, never simply identification. He says short-answer questions would also work with his technique, but not essay questions, as far as he can tell, because one student would most likely end up writing the entire exam.

Although Rab's exams are 50 percent group and 50 percent individual, and the group test portion of the final grade is 40 percent, some students complain that his method isn't fair. They ask, "What if I do all the work and someone else gets the same grade?"

Rab's goal is to have group members design and enforce certain rules to keep everyone accountable. This means fairness becomes the students' responsibility. By the end of the quarter, most students have come to find the exercise instructive and fair, especially because the instructor rewards improvement and occasionally eliminates a "bad day" on individual exams. Honors students tend to be particularly good cooperative learners, according to Rab, who takes pains to assign students randomly to groups of three or four based on availability to meet for at least 1 hour per week outside of class. In time, all of his students improve their cooperative work skills.

Although he has only 12–15 students in his class, Rab believes his system could work with more. A colleague of his, a microbiology instructor, uses the same technique with 35 students in groups of 5 or 6.

OPEN-BOOK, POSTEXAM QUESTION CHALLENGES

Marilyn Shopper, Science, Health Care, and Math Division, Johnson County Community College, 12345 College, Overland Park, KS 66210; TEL: (913) 469-8500 Ext. 3387; FAX: (913) 469-2518;
E-MAIL: mshopper@johnco.cc.ks.us.

Courses Taught

- Principles of Biology (and lab)

Description of Examination Innovation

About 4 years ago, Marilyn Shopper began making postexam review sessions in biology more productive learning experiences for her students.

After grading and returning exams, which usually consist of approximately 40 multiple-choice questions and 5–10 short-answer questions, Shopper allows her students 20–30 minutes to challenge, criticize, or comment on any of the questions, using their textbooks and notes to explain why they believe their answer is correct. Students usually focus on the multiple-choice questions, justifying, debating, and posing questions such as whether a certain answer is *always* true. "Some are just looking for a point," Shopper says, "but others really want to understand why an answer is correct or incorrect."

The first time she does this exercise with a new class, students are tentative. It takes one run-through, says Shopper, for them to gain confidence in the process and in themselves.

Early in her use of this technique, Shopper recognized an error in her answer key just before she was to return graded exams. Instead of correcting it, she challenged her students to find the mistake during the posttest review. She told them that the only way they would get credit for their answer was if they found the error. Given this incentive, the error wasn't too difficult for her students to find. "Now I purposely make mistakes on the key and challenge the students to find the errors," says Shopper.

Sometimes students pose answers similar enough to the correct ones that Shopper agrees to give additional credit. She says the reviews do not become gripe sessions, as students know "they have to have a good argument to earn the points."

Her method has real appeal to students once they get used to it. Some have told her that "going over the exam" is their favorite time in class.

NARRATIVE COMPLETION PROBLEMS

Gil Starks, Department of Biology, Central Michigan University, Mt. Pleasant, MI 48859; TEL: (517) 774-3607; FAX: (517) 774-3462; E-MAIL: g.starks@cmich.edu.

Courses Taught

- General Biology
- General Botany

Description of Examination Innovation

In principle, essay exams would be the best way to test students' understanding of science, says biology professor Gil Starks. But in large classes, this is difficult to do. Instead, in his general botany courses of 160–200 students, Gil Starks's exams employ a narrative format for "completion" problems. The format, he believes, encourages students to synthesize material—to see the forest through the trees—while at the same time being machine scorable, like multiple-choice test items. But unlike mutiple-choice exams, Starks says his completion problems will encourage students not to learn only atomized bits of knowledge, but, rather, the power of coherent statements.

Starks' completion problems are not limited to sentences. Students in Starks's classes are sometimes given a one- or two-paragraph discussion of a topic and told to fill in blanks with terms supplied on the exam. Students are supposed to develop a "flow of ideas" during the test instead of supplying single-statement answers.

For example, a section from one of Starks' exams provides two paragraphs and four terms from which to fill in the blanks:

> Classification of organisms has been done in several ways by several people over the past several years. Now biologists generally agree that all living things can be organized under five _____.
>
> Classification of living things is done by grouping together those that have features in _____.
>
> Choices: (A) classifying; (B) family; (C) kingdoms; (D) species.

Starks concedes that some students unduly benefit from the context of the discussion and may occasionally put in a correct word without entirely understanding the concept. He is working at eliminating most "giveaway" answers. But he suggests that divining an answer from a given context may not be as much of a problem as it first appeared. "Many students don't read or synthesize all that well when they start college. With the 'story' format, I can start to teach them how to do this," says Starks. He believes tests like these help students make connections. They get a *feeling* for synthesis while taking a test, which turns the exam into a teaching tool.

UNUSUAL QUESTION FORMAT

Lloyd L. Willis, Department of Biology, Piedmont Virginia Community College, Route 6, Box 1, Charlottesville, VA 22901; TEL: (804) 977-3900; FAX: (804) 296-8395; E-MAIL: lw@jade.pvcc.cc.va.us.

Courses Taught

- General Biology
- Biological Problems in Contemporary Society

Description of Examination Innovation

Lloyd Willis uses unusual and creative test formats to ensure that his students are not merely memorizing biology but also understanding the subject.

For example, in his famous "BB" question, students must explain *digestion*, *absorption*, and *ingestion* as they relate to the presence of BBs, the kind shot from BB guns—objects not typically associated with the digestive system. Two BBs, students are told, one plastic coated and one uncoated, are present in some cookie dough that is about to be eaten. Using the three key concepts, students must describe what happens to each BB as it is consumed. They are reminded to consider the pH of the stomach.

There is usually one question on each of Willis's tests that generates a great deal of student discussion in the hall after the exam. This question gets the "prize" in the digestive tract unit. The hints given have varied from test to test, reminding students of stomach secretions, acid, the gall bladder, and the like. Willis says that if a student can specifically state what is happening to each BB as it passes along the digestive tract, he/she understands the various functions of the system.

"This is a more complicated version of the question of what happens when the 2-year-old eats the sand in his/her playbox," says Willis.

In another innovative test question, students are asked to write "chemical formulas" based on being given four coins, each with a different value and on an assigned electrical charge. The question is worded as follows:

> Place "chemical formulas" on the left based on the information that follows. If the coins below had the following symbols, value, and electrical charges, write one or more formulas that would represent the concept of a chemical compound. The value for each coin is a factor in the answer, as well as the electrical charge.
>
> Symbols for the coins and their values are given under the question: 50-cent piece = +50, quarter = +25, nickel = −5, dime = −10.

If the students understand the concepts underlying chemical formulas, they can apply that information to the new symbol and the assigned electrical charge. If the students have memorized a chemical formula (such as NaCl) and do not understand why it is written this way, the question will be too difficult. Over 70 percent of Willis's students answer the question correctly, and about 10 percent do not attempt it. Challenging questions such as these encourage students to go beyond recall. Faculty colleagues also appreciate questions that show students understand what they have learned.

CHAPTER 4

CHEMISTRY

CONCEPTUAL EXAMINATION

American Chemical Society, Exams Institute, 223 Brackett Hall, Clemson University, Clemson, SC 29634; TEL: (864) 656-1249; FAX: (864) 656-1250; E-MAIL: acsxm@clemson.edu *or* http://tigerched@clemson.edu.

Description of Examination Innovation

The efforts of the American Chemical Society (ACS) to reform the first-year general chemistry course were enhanced in 1994–1995 by the development of a conceptual examination for the course, prepared by a committee (chaired by Diane Bunce of the Catholic University of America) of well-known undergraduate chemistry education reformers at ACS's Examinations Institute at Clemson University. Their purpose was to see whether there is a way to test conceptual understanding within the tradtitional multiple-choice format, a format that lends itself to machine grading in large classes and allows national statistics to be gathered. They began their work by devising some criteria for "conceptual understanding"—criteria that are well represeted in this collection. For the ACS group, conceptual understanding involves the "integration of one or more concepts with other knowledge held by the learner as applied to a problem or a question previously unseen."

The group proceeded to develop questions by starting with traditional general chemistry examination questions from which in each instance the underlying concept was identified. The traditional question was intended to test understanding of the concept, but often, because students could answer such questions using factual recall or memorized algorithms, it did not. Once the key concept was identified, the group rewrote the question stressing the explanation (the why) of the chemical phenomenon presented. Specifically, they were looking for questions to serve as "occasions of learning as well as evaluation."

The model exam the ACS is now distributing is meant to stimulate other chemistry faculty to rethink their own exam practices and to come up with questions and formats that reach the conceptual core of chemistry. Within a "tiered multiple-choice format," the new ACS exam tests students on one or more of the following competencies: ability to visualize a system and use it to reach a conclusion; ability to predict from a given chemical situation what might happen next, why, and how; ability to think about a chemical situation on the atomic/particulate level; and the ability to select from a "grid" holding many partial answers all possible answers to explain a phenomenon.

The ACS conceptual exam consists of 60 items in a standardized multiple-choice format and takes 110 minutes to complete. It is available to faculty from the ACS Exams Institute at the address above.

HANDS-ON LABORATORY ACTIVITIES ON EXAMS

Jerry A. Bell, formerly of Simmons College, American Association for the Advancement of Science, 1200 New York Avenue, N.W., Washington, DC 20005; TEL: (202) 326-6786; FAX: (202) 371-9849; E-MAIL: jbell@aaas.org.

Courses Taught

- Introductory Chemistry
- General Chemistry
- Introductory Organic Chemistry
- Introductory Biochemistry
- Physical Chemistry
- In-service workshops for elementary, secondary, and college teachers of science (emphasizing chemistry and inquiry)

Description of Examination Innovation

About 5 years ago, when Jerry Bell taught introductory chemistry at Simmons College in Boston, he wanted to make his exams "more like real chemistry, testing students on more of the work chemists really do." He was particularly interested in having students observe chemical reactions and then be able to interpret what they observed in scientific terms.

Not that introductory chemistry students at Simmons didn't already have 3-hour laboratory exams measuring their effectiveness in the lab. But Bell wanted to integrate the skills they learned in lab into his exams, so he started including hands-on laboratory activities on his tests.

The laboratory portion of Bell's exams in no way supplants the longer laboratory practical—the activities required of students are simple, such as mixing two solutions together and observing, or adding one solution to another, drop by drop. What is most important to this instructor is that students record their observations and make interpretations based on the phenomena they observe, something Bell feels introductory chemistry students often are uncomfortable doing.

In fact, Bell first devised this innovation to see whether there was a difference in students' ability to interpret phenomena when they were told about the phenomena as contrasted with observing the same phenomena for themselves. He suspected that students generally do not trust their own observations. "If you or the textbook tells them that a white precipitate is formed, then they are able to tell you why—provided they've truly understood the chemistry. But, when they see it forming in front of their eyes, they become hesitant. They first have to decide if they're actually seeing a white precipitate," Bell explains. "Even when they're pretty sure they have the 'right' observations, they still have to translate those observations into an explanation or interpretation." Bell says he would have to do a longer comparative study to report his "results" definitively, but the fact that his students performed about 10 percent more poorly on the hands-on portion of these exams than on the standard paper-and-pencil portions suggests they do lack confidence in their ability to make appropriate observations.

Bell's students take the lab portion of his final examination in class and do the midterm lab section on their own time in the laboratory. They use microscale materials—plastic pipettes filled with chemical solutions, plastic well plates, acetate sheets, and small test tubes. Microscale technology has made Bell's innovation much easier to do in the classroom under timed conditions.

Open-ended exam questions come in one of two general forms: in one, students are told to mix specific known reagents, to observe the results, and to figure out "what's happening"—that is, to explain the phenomenon they've observed from a chemical point of view. In the other, students are *not* told the nature of the reagents they're mixing. They make observations and hypothesize what reagents, or class of reagents, they were given, based on the results and their knowledge of chemistry.

Is Bell concerned, during out-of-class exams, about students talking to each other before turning in their papers? The instructor says "no." He finds that most students are honest (if they expect to be trusted), and, after years of writing open-ended questions, he knows what to ask, so students really can't cheat without direct plagiarism, which is easy to detect. There isn't a single right answer to his questions, and there isn't a "book answer" to his problems, either.

As a result, when grading these questions, Bell looks at more than the answer. Some credit is given for "correct" observations, to verify that students have carried out the procedure properly. But the instructor looks harder at their interpretation—even if their observations aren't "correct" (though entirely incorrect observations have been made only once out of about 200 cases). He's looking for interpretations that are consistent with the results. Perhaps the student saw only one white precipitate, for example. If there should have been two, some points are taken off, but partial credit is given if the student can reason logically based on the evidence.

"I give lots and lots of partial credit on the standard paper-and-pencil questions, too," says Bell, who grades all papers one question at a time to guarantee consistency.

"By the tenth exam, you've seen many of the most common student problems," he explains. "You learn what students don't understand, and you always have the option to go back and regrade the most problematic items if you want." Bell says he has "almost never" given a multiple-choice or true–false question on an exam.

When asked if he's considered using cooperative groups for this portion of the exam, Bell says he has done so only with high school teachers in in-service workshops. The group format does create "a richer variety of explanations" for the observed phenomenon, and if he were still teaching, he says he would use groups more extensively.

Students seemed to enjoy the lab exams, and colleagues were intrigued with the idea, too, though Bell doesn't know if any have gone on to use laboratory activities on examinations.

GROUP QUIZ BASED ON IN-CLASS DEMONSTRATION

Craig W. Bowen, Department of Chemistry and Biochemistry, University of Southern Mississippi, Box 5043, Hattiesburg, MS 39406-5043; TEL: (601) 266-6044; FAX: (601) 266-6075; E-MAIL: cbowen@wave.st.usm.edu.

Courses Taught

- Chemistry for nonmajors (at Purdue University)
- General Chemistry for science and engineering majors

Description of Examination Innovation

By focusing on in-class demonstrations, Craig Bowen, an assistant professor at the University of Southern Mississippi, has found a way to include complex experimental chemistry problems on in-class exams, problems like those encountered in the laboratory, without limiting the range of material assessed. The "demonstration-based group quizzes" he began using while team-teaching at Purdue (and later used at Florida State University and the University of Southern

Mississippi) also help students learn to think more critically and analytically about in-class demonstrations.[1]

After observing many chemistry, physics, biology, and geology classes, Bowen says students do not learn well from demonstrations:

> Often students do not take notes during the demonstrations, and do not summarize what the demonstration actually demonstrated. When they're asked, "What did the demonstration actually show?" they usually respond, "Gee, I don't know, but it was neat!" By making some portion of assessment based on demonstrations, I am trying to get students to ask themselves, "Now, what does this demonstration imply about the concepts that have been presented?"

When Bowen and Phelps started to use demonstration-based group quizzes, they were given approximately 2 weeks before each exam (three per term) and covered material discussed or demonstrated in the period since the previous exam. The instructor gave a demonstration while students recorded the data on sheets that were part of their quiz. They then answered approximately four questions about the results of the demonstration.

For example, one demonstration-based group quiz involves a multistep titration. Students are given the concentration of a sodium hydroxide solution. Next, phenolphthalein is added to two flasks marked A and B. Flask A contains a solution of nitric acid, and flask B contains sulfuric acid. Each flask of acid is titrated to the end-point with the sodium hydroxide solution, and the students record the data in their table. After the titrations, a small amount of barium nitrate solution is added to each titration flask. In flask B, a white precipitate forms, whereas there is no visible reaction in flask A. Students are asked to gather the following information during the demonstration:

- Concentration of sodium hydroxide solution.
- Amount of sodium hydroxide solution needed to neutralize the acid in flask A.
- Amount of sodium hydroxide needed to neutralize the acid in flask B.
- Result of reaction with barium nitrate and remains of neutralization in flask A.
- Result of reaction with barium nitrate and remains of neutralization in flask B.

Students then answer questions about the demonstration, such as to determine (1) the identity and concentrations of acids A and B and explain how they arrived at their results; and (2) the identity of the solid formed in flask B and predict how much was formed.

---

[1] Bowen developed this technique while a graduate assistant at Purdue University with Amy Phelps, who presently teaches chemistry at the University of Northern Iowa.

In an open-ended questionnaire designed to gain feedback about the course, Bowen discovered that some students had problems with the collaborative aspect of the quizzes. A student indicated that he felt it was unfair, because "less smart people sit next to me and get an A on their quiz because of the work I did, or vice versa." Some complain that group work takes more time than might be expected, and that Bowen should allow more than 20 minutes for them to complete the quizzes. Others asked for more quizzes, and with questions that are more straightforward and independent of one another than are Bowen's demonstration-based quiz questions. One student wrote, "It is debilitating to not get parts two, three, and four right, even if you knew how to do them, simply because you didn't know how to do part one."

Based on the feedback from students, Bowen has changed his approach to using demonstration-based group quizzes, in addition to increasing the frequency from three to five per term. To address the issue of time pressure raised by students at Florida State, he now starts the demonstration-based assessment task during lecture 2 days prior to the written exam. Students are given a sheet of questions related to the demonstration that they turn in as part of the written exam they take in class 2 days later. The demonstration questions are worth 10 points of the entire 100-point exam, and each question (with usually three to five parts) can receive partial credit. This 2-day period provides students an opportunity to work together outside of class so time is less of an issue.

Surveys of students in the demonstration-based class compared to a control group indicate that the students in the demonstration-based class study about one and a half hours per week with other students in the class, compared with the 45-minute average that the nondemonstration students spend working together outside of class. In addition, analysis of common exam items from the final exam indicates that in those content areas in which students have demonstration-based assessment throughout the term, students score significantly higher than their colleagues in the nondemonstration class. In those content areas in which demonstration-tasks are not provided, there is no difference in student performance on the final exam items. Because of this demonstration of the effect assessment practices can have on student comprehension, some of Bowen's colleagues are considering incorporating these practices into their sections of general chemistry.

Complications do arise in demonstration-based assessment: The demonstration fails to work properly, the demonstration takes too long to set up and perform, or students are unable to see the demonstration. However, Bowen says most of these problems can be overcome. He advises his colleagues to choose *simple* demonstrations to perform that do not require elaborate setups, and to try the demonstration more than once to make sure it works. An alternative to live demonstrations is to use a videotaped demonstration or videodisc, which can be shown several times, allowing students to make appropriate observations. Finally, student-performed demonstrations permit students to see better and to manipulate some of the variables themselves as they answer an instructor's questions.

COMBINATION MULTIPLE-CHOICE/OPEN-ENDED QUESTIONS, GRADED GROUP DISCUSSION, AND HELP DURING EXAMS

Diane M. Bunce, Chemistry Department, The Catholic University of America, Washington, DC 20064; TEL: (202) 319-5390; FAX: (202) 319-5381; E-MAIL: bunce@aol.com.

Courses Taught

- Chemistry 101—Chemistry for Health Professionals (Nursing Chemistry)
- Chemistry 125—Chemistry in our Lives (nonscience majors)
- Chemistry 126—Chemistry in the World around Us (nonscience majors)

Description of Examination Innovation

Over the last 10 years, Diane Bunce has used some innovative testing practices that not only enhance the feedback she gets from her exams but also encourage her students to become more comfortable writing and talking about science. In her Chemistry 101 course, she uses open-ended questions *in combination* with multiple-choice test items to probe students' understanding of the concepts of chemistry that underlie test questions. For example, after answering a multiple-choice question requiring identification of the molar solution with the highest osmotic pressure, the student must go on to explain or defend that answer in the short open-ended question that follows. With this method, students cannot rely on memorizing their notes to pass the test. They are forced to focus on understanding the chemistry.

Bunce also uses essay questions on chemistry exams for her nonscience majors to get them to articulate their views on controversial scientific issues. Here, too, students take a position, logically explain it, and then defend their choice. An example of an essay question she has used in this course describes a television ad for Bayer aspirin, which states that when asked, nine out of ten doctors agreed that if stranded on a desert island, they would choose Bayer aspirin over its competitors as their drug of choice. Students are asked to describe what is scientifically correct and/or scientifically misleading about the ad. Bunce says she grades these essays on the quality, quantity, and accuracy of the science presented and on the logic of the students' arguments.

To further develop students' critical thinking about scientifically based social issues, she organizes her classes into discussion groups, usually of four students, and then presents each group with an issue to discuss from a scientific, political, economic, and ecological point of view.[2] A discussion summary sheet is filled out

[2] As a member of the team that developed and incorporated this approach into the ACS curriculum *Chemistry in Context* (Dubuque, IA: Wm. C. Brown, 1994), Bunce feels nonscience majors have unique contributions to make to their chemistry studies because of, not in spite of, their other majors, and that activities such as these capitalize on their evolving expertise.

together by the students during their discussion. The sheets describe the topic to be discussed and, where necessary, provide additional background information. The topic in question is then broken down into shorter, more direct questions that help the group focus on the important issues. The last part of the sheet is typically a request for a letter to a legislator or administrative official, in which students must give their view on the issue and defend it using information already worked out in the shorter questions. Wherever possible, students are asked to role-play as they write the letter together, and in at least one section of the sheet, students are required to detail the science involved in the issue. Bunce and her TAs grade each group on the quality of the completed sheet, giving every student in the group the same grade.

The discussion summary sheets started out for Bunce as a way to keep students on task; they also provided the instructor with a way to hold students accountable for their work. Now Bunce sees that students view them as a way to focus in on the issue at hand. Students, too, say they like having the opportunity to hear what others in the class think about the issues covered.

Finally, Bunce is trying to decrease test anxiety in her classrooms by circulating among her students while they are taking an exam and offering help and encouragement. If a student raises a hand and asks for help, often Bunce will take a seat nearby and listen. Sometimes students want to "try out their own logic" for an answer; sometimes they "go blank" and need help to get back on track. And if a student is having trouble understanding a question, Bunce will assist.

Bunce says her role during in-class examinations is to help the students stay calm by asking short questions that direct them to a successful solution. When students are too lost for help, she will tell them that they do not have the requisite knowledge to answer the question and will invite them to come to her during office hours so they can analyze together the students' study habits.

SETTING CLASS AVERAGE TO 70 PERCENT OF MAXIMUM POSSIBLE SCORE

John F. Cannon, Department of Chemistry and Biochemistry, Brigham Young University, C161 BNSN, P.O. Box 25700, Provo, UT 84602-5700; TEL: (801) 378-3301; FAX: (801) 378-5474; E-MAIL: jfcannon@chemgate.byu.edu.

Courses Taught

- Introductory Chemistry

Description of Examination Innovation

For the past 9 years, John Cannon has used a combination of grading techniques that he says "essentially eliminate competition among students" in his introductory chemistry classes.

His method is as follows: Cannon makes adjustments to exam and quiz grades so the class average will equal approximately 70 percent of the maximum possible score. This is accomplished by multiplying each student's score by the ratio of the maximum possible score to the highest student score. A number is then added to (or subtracted from) each score to make the average score equal to about 70 percent of the maximum possible score. For scores above the average, this is done by a sliding scale, designed to prevent any score from exceeding the maximum possible score. This adjustment places each evaluation on an equal footing, regardless of its degree of difficulty (e.g., a score of 57 on an exam with an average score of 54 is better than a score of 72 on an exam with an average score of 77.)[3] The adjustment process also involves fixing the sum of exams, sum of quizzes, and the final exam, at their appropriate percentages to arrive at an adjusted total score for each student.

Cannon makes adjustments in assigning final course grades as well. He awards an A to all students within 4.5 percent of the top student's adjusted total score. Ranges for A– and B+ are also 4.5 percent. A range of 5.5 percent applies to B, B–, C+, C, and C– grades, and of 3 percent for D+, D, and D– grades. Anyone scoring less than half the score of the top student receives a failing grade. If this scale results in a course grade point average (GPA) less than 2.3, Cannon relaxes the scale enough to raise the GPA to 2.3. If applying this scale results in a course GPA exceeding 2.3, then "so be it," Cannon says. If there is an exceptional student in the class who far outscores all the others, Cannon will base grades on the second student.

Students are told that final exam grades carry extra weight in the determination of course grades. However, no one's course grade, Cannon promises, will differ by more than one full grade from their final exam grade, *regardless of the student's final adjusted total score.*

Cannon feels these methods eliminate the perceived *arbitrariness* out of grading curves: "Everyone within 4.5 percent of the top student gets an A, regardless of how many there are." Cannon says that generally, faculty do not agree on a definition for grading on a curve. Although his class's grade distribution includes a "considerable bulge" in the high-C or low-B range, the shape of his curve is very much determined by the students' performance and is not consistent year after year. Upon review of the distribution for the last ten Chem 106 classes he taught, Cannon notes that there is a "large variation," even though the class G.P.A. has a fairly narrow range of 2.36 to 2.55.

It never occurred to Cannon to investigate student reactions to his grading system until he reviewed a recent set of student evaluations. He discovered no positive

---

[3] His formula for this alteration is: $NS = SS(MS/HS)$ and $AB = 0.7(MS) - AVE$.
Definitions: MS—maximum possible score; HS—highest student raw score; SS—student raw score; AB—adjustment base; NS—normalized student score; AVE—average normalized student score; AS—adjusted student score. For each student at the average or below, $AS = NS + AB$. For each student above the average, $AS = NS + AB - [(NS - AVE)/(MS - AVE)]$. The term subtracted here ensures that no score exceeds MS.

comments, and several "very negative" ones, about his grading scheme. Although most of the negative comments revealed a lack of understanding of the grading system, Cannon decided to offer students a choice of grading methods: the modified curving system described above or a rigid curve with a class GPA equal to the average of his last ten Chemistry 106 classes. Both systems were explained in the syllabus and during the introductory lecture of his winter 1996 Chemistry 106 class. Of 169 students, 147 opted for Cannon's flexible grading system during the first week of the semester. Of the 22 students who chose the structured system, 2 received higher and 2 received lower grades than they would have received in the flexible system. Of the 147 students who chose the flexible system, 18 received higher and 10 received lower grades than they would have received in the structured system.

Cannon knew, if given this choice, that students would make an effort to understand both grading schemes. And negative student evaluations of the fairness of Cannon's grading disappeared. One student even wrote, "Grading policy (structured vs. flexible) was most fair I have ever seen."

Although not *all* competition is eliminated with Cannon's flexible grading system, what students probably perceive as the most unfair aspect of the normed curve—competing for a certain number of A's, B's, or C's that the teacher has allotted before the class has begun—doesn't exist. Here, students are being provided with a system that may make more sense to them.

MIXED-FORMAT EXAMS

Lyman R. Caswell, Professor Ementus, Department of Chemistry and Physics, Texas Woman's University, P.O. Box 425859, Denton, TX 76204-3859; TEL: (817) 898-2550; FAX: (817) 898-2548.

Courses Taught

- Chemical Principles I and II

Description of Examination Innovation

Lyman Caswell's experiments with mixed-format exams have made him aware of how strongly testing affects student learning.

For a long time he was dissatisfied with multiple-choice tests, feeling they tested students' ability to do multiple choice—that is, their *facility* more than their understanding. When he became an instructor in the honors section of freshman chemistry in 1988, Caswell was able to break from the departmental "multiple guess-whats" favored by most of his colleagues. For 2 years, he gave mixed-mode exams, with problems to solve, fill-in-the-blanks, some essay questions, and a short multiple-choice section. Many of the students who did poorly on the multiple choice did well on the other portions of the tests. By 1990, he stopped using multiple-choice questions altogether and increased the number of problems and essay questions.

By 1992, however, it was clear to Caswell that by the time they came to his course, most of the self-selected students in the honors sections had learned to excel in science courses through solving numerical problems. Absent the math difficulties that plague average chemistry students, typical honors performance on problems ranged from 80 to 85 percent. However, these same students averaged only 45–50 percent on essay questions. The essay questions did not involve lengthy written answers. Typically, the test would include questions such as why dissolving sodium chloride in water does not involve a chemical change, or, concerning periodic law, why the tendency of atoms to gain electrons increases from left to right in the periodic table. Caswell says he had to tell students that for questions beginning with the word "explain," a simple restatement in different words would be an inadequate response.

In defense of his essay questions, Caswell says in all cases the concept was discussed in class before being presented on the test. Furthermore, the questions he wanted to ask dealt with concepts that were not easily tested with numerical problems, at least not at the freshman level.

In discussions with other readers of essays on the Chemistry Advanced Placement test of the Educational Testing Service, Caswell found that his experiences were not unique.

OPEN-ENDED QUESTIONS, LIBRARY RESEARCH, INCORPORATING IMPROVEMENT INTO COURSE GRADE

Brian P. Coppola, Department of Chemistry, University of Michigan, Ann Arbor, MI 48109-1055; TEL: (313) 764-7329; FAX: (313) 747-4865; E-MAIL: bcoppola@umich.edu.

Courses Taught

- Structure and Reactivity

Description of Examination Innovation[4]

To achieve their goals for their new organic chemistry–based introductory course, Brian Coppola and Seyhan Ege, cofounders of the University of Michi-

[4] What follows is adapted from B. P. Coppola, "Progress in Practice: Using Concepts from Motivational and Self-regulated Learning Research to Improve Chemistry Instruction", in P. R. Pintrich, ed., *Understanding Self-regulated Learning: New Directions in Learning and Teaching*, no. 63 (San Francisco: Jossey-Bass, 1995), pp 87–96. For a detailed description of the course in question, see Sheila Tobias, *Revitalizing Undergraduate Science: Why Some Things Work and Most Don't* (Tucson, AZ: Research Corporation, 1992), Chapter 4.

gan course, set out to correct what they think are some wrong messages that students take away with them from traditional examinations in chemistry, and to convince them, instead, of the following: (1) that memorizing chemistry facts and examples is not enough; (2) it is important to understand what chemists actually say about the things they study; and (3) the same phenomenon can have multiple representations (words, pictures, graphical and numerical versions).

The exams for Structure and Reactivity, taken by 1200 students at any one time, are designed to achieve these learning objectives. Some questions are intentionally designed to yield many correct solutions within the context of the course and given the information provided. In some instances, the type of question simply lends itself to multiple answers, such as with synthesis sequences; in others, multiple interpretations are what any chemist would propose. On nearly every exam, students are encouraged to suggest reasonable alternative solutions (some that the instructors themselves did not anticipate).

Coppola and Ege accomplish their course goals by relying on primary literature for the examples that appear on their exams. Along with a citation, the impact of using clearly contemporary examples communicates to students that strategies focusing on recognition as the sole outcome will be unsuccessful. With no chance that the particular example will have been encountered, the students must actively face the difference between a general concept ("the forest") and the many representative examples ("the trees"). In an extensive course packet containing both learning advice and previous exams, the message is clear: On the 1994 exams, the citations are from 1994, and so on.

Another consideration in designing exams for the new course is the instructors' acknowledgment that students develop new skills at different rates. Since the course is intended to be cumulative, they have devised ways to make improvement count. One technique is to increase the point value of exams throughout the term (without increasing the length of the exam). Thus, it is "worth" more to do better later. The instructors also make judgments about improvement by considering the set of exams and the final as two independent measures of cumulative performance.

The instructors take exams as seriously as their students—but for different reasons. "A set of examinations outlines the expectations of a course much better than a syllabus," writes Brian Coppola in the introduction for a collection of examinations used in Structure and Reactivity.[5] He continues: "If these goals

---

[5] Brian P. Coppola, Seyhan N. Ege, and Richard G. Lawton, "The University of Michigan Undergraduate Chemistry Curriculum, Part II: Instructional Strategies and Assessment," *Journal of Chemical Education* 74(January 1997): 84. (in press). Copies of the current collection of examinations and student learning tips for the Structure and Reactivity courses (Chemistry 210 and 215) can be obtained from Hayden-McNeil Publishing, 47461 Clipper Road, Plymouth, MI 48170; Tel: 313–455-7900, FAX: 313-455-3901.

also include higher-order learning and thinking skills, then care must be taken to preclude [the development] of unwanted skills." In other words, if the instructor does not want students to memorize and regurgitate, then he/she must design tasks that do not reinforce these skills and include explicit instruction for alternative strategies.

In the second term of Structure and Reactivity, there is a special section of 120 self-selected students who desire a research-oriented instructional environment. The final exam in this section consists of photocopies of four to five short research publications and an accompanying set of leading questions. To prepare students for this sort of examination, weekly library exercises are assigned, along with peer-group discussion sessions. Students take a general idea from the course and then search the primary literature to find examples of it. In one exercise, students create examination questions the way their instructors do. Different kinds of literature-searching techniques are integrated into the exercises throughout the term. "Students," Coppola writes, "are constantly amazed by the fact they can find meaning in state-of-the-art information—a skill which is one of the goals the instructors have for the course."

Students appreciate their more imaginative exams. A typical response is this one: "What a pleasant experience taking this exam was. For the first time, I felt as though the exam was an accurate reflection of what [I] was supposed to get out of the combination of the lecture and the text."

CROSSWORD PUZZLES

Charles H. Corwin, Department of Chemistry, American River College, Sacramento, CA 95841; TEL: (916) 484-8474; FAX: (916) 484-8674;
E-MAIL: chcorwin@aol.com.

Courses Taught

- Introductory Chemistry

Description of Examination Innovation

Dissatisfied with the examinations he previously employed, which focused on key terms and definitions, in 1994 Charles Corwin began to experiment with computer-generated crossword puzzles that utilize key terminology.

The results of this testing method show quantifiably improved scores. Perhaps more important, students appear to enjoy working with the puzzles and have communicated to the instructor that they feel more comfortable with the language of chemistry as a result. Another spinoff, reports Corwin, is that spelling of technical terms has improved, especially for multicultural classes.

See the following example.

Chapter 5 Key Terms

Across

1. the distance light travels in one cycle
3. a spectrum with broad bands of radiant energy
7. an electron orbit designated 1, 2, 3 ... (2 words)
8. the arrangement of electrons in an atom
10. a spectrum with narrow bands of radiant energy
14. a region of high density in the center of the atom
16. a model of an atom with electron probability (2 words)
20. the ______ number refers to the *A* value
23. a formula for a wavelength of light from a H atom
24. a model of an atom with electron orbits
25. a neutral subatomic particle
26. the number of wave cycles in one second

Down

2. the light spectrum from 400–700 nm
4. an instrument for determining the mass of an isotope
5. a region of high probability for finding an electron
6. an expression for the composition of a nucleus (2 words)
9. 1/12 the mass of a carbon-12 atom
11. atoms that differ only by the number of neutrons
12. the ______ number refers to the *Z* value
13. the spectrum from x-rays through microwaves (2 words)
15. a particle of radiant energy emitted from an excited atom
17. the principle that the precise location and energy of an electron cannot be determined simultaneously
18. a negatively-charged subatomic particle
19. the weighted average mass of isotopes (2 words)
21. an energy level that is designated *s*, *p*, *d*, *f* ...
22. a positively-charged subatomic particle

Corwin uses the Macintosh application *Crossword Creator* to generate puzzles relatively easily and usually in less than 30 minutes. If sent a self-addressed stamped envelope, Corwin will provide a set of puzzles (20 puzzles, each with approximately 25 key terms) for introductory chemistry. The puzzle terms are arranged by topic to accompany the introductory textbook *Chemistry: Concepts and Connections* (Prentice Hall, 1994).

MODIFIED PERSONALIZED SYSTEM OF INSTRUCTION/MASTERY LEARNING

Mark S. Cracolice, Department of Chemistry, The University of Montana, Missoula, MT 59812-1006; TEL: (406) 243-4475; FAX: (406) 243-4227; E-MAIL: markc@selway.umt.edu.

Courses Taught

- Preparation for Chemistry

Description of Examination Innovation

In his Preparation for Chemistry course, for students who feel inadequately prepared for general chemistry, Mark Cracolice uses a modified Personalized System of Instruction (PSI)/Mastery Learning instructional design. This system uses the traditional PSI reinforcement, cues and feedback, and mastery learning (three of five influences H. J. Walberg identified, in a 1984 synthesis of 3000 studies, as producing the greatest positive effect on student achievement).[6]

Cracolice's course is divided into 19 self-contained units of study. Assignments for each unit are given in the course syllabus, and students study for each unit assignment at their own pace, subject to grade penalties if they do not meet deadline dates. Each student has three attempts to "master" the material in each unit, as demonstrated on a short unit test. Some unit exams are all multiple-choice questions (used for theory), some are all open-ended (used for calculational problems), and some are mixed. A score of approximately 90 percent is required to pass a unit. Students are given unlimited time to complete the test within time limits of the tutoring sessions (2–4 hours per day).

Each sequential version of the unit tests is slightly different from the other forms. For example, one test will ask students for the temperature of a gas at a given pressure and volume. Another will ask them to find the volume given a different pressure and temperature.

Grades for the course are determined by two midterms and a comprehensive final exam. Although a fair number of students passed the midterm and final in one recent year, a large portion (43 percent of the original enrollment of the

[6] The other two were reading training and acceleration. In H. J. Walberg, "Improving the productivity of America's schools," *Educational Leadership* 41, no. 8(1984): 19–27.

class) stopped attending (dropped, withdrew, or failed without taking the final) during the course of the semester.

Cracolice plans to further modify the "self-paced" aspect of his course. Because most students only take the tests when they are pressured by deadline dates, he will increase the number of unit tests to 43, each covering less material. He will also require students to sign up for unit tests (first-time or retests) at least three times a week to get students into a habit of regular study. Material in each textbook chapter will be covered in approximately two and a half unit tests (instead of one), and following each chapter, a comprehensive unit test will be administered covering all the prior material, with an emphasis on the more recent material. This last modification occurred to Cracolice after noticing that a significant source of difficulty on any given unit represented the student's failure to have truly mastered information from a previous unit. (Note the percentage of A's and B's reversed from midterm to final grades.)

As an additional motivational tool, Cracolice plans to have students keep a log of the hours they spend working on his course. He feels many students do not spend enough time studying for any course, not just PSI. The purpose of the log is to make students realize that one source of their difficulties is inadequate study.

In end-of-course evaluations, students who successfully completed Cracolice's course indicated that the advantage of this system is that the student "must know the material," to quote one student. Others wrote that the course makes them take responsibility for their grades, and this carries over to other classes: "This class... has changed the way that I do things. I now actually study during the week, even when I don't have any tests," reported one student. Criticisms included the loss of the weekly lecture and the time required to do the course work.

PARTIAL-CREDIT MULTIPLE-CHOICE QUESTIONS

Arnold Craig, Department of Chemistry and Biochemistry, Montana State University, Bozeman, MT 58717; TEL: (406) 994-4801; FAX: (406) 994-5407; E-MAIL: uchac@earth.oscs.montana.edu.

Courses Taught

- General Chemistry

Description of Examination Innovation

In 1989, Arnold Craig, Reed Howald, and colleagues teaching general chemistry at Bozeman substituted partial-credit multiple-choice questions on examinations in place of standard TA-graded quantitative problems in an attempt to find a way to give partial credit while using machine-scorable multiple-choice exams. The reason they switched to the partial-credit model is that they found themselves for the first time in their teaching careers team-teaching a single course. In a team-teaching environment, they were frustrated by the lack of consistency in their own or their TAs' hand-grading of their exams, which were half quantitative problems, half qualitative questions requiring a variety of answer formats (multiple choice; matching lists; one-word or short, written responses).

Instead of selecting the standard multiple-choice format with four or five answer choices, the team invented their own partial-credit multiple-choice format that involved the following change. Students are offered 10, sometimes 15, choices, of which 3 or 4 may be "correct enough" to be given credit. The maximum score assigned to a question will be variable, but is usually around 4 or 5 for nonnumerical answers, 7 for numerical answer problems. The team found it easy to give partial credit for not-quite-perfect answers on both qualitative and quantitative questions.

For example:

> "Professor Howald just ordered 1 cubic yard (1 yd^3) of crushed stone for his driveway. How many cubic inches (in^3) did he receive?"

> Ten distracters are given. The correct answer, "4.67×10^4," is worth 4 points. "1296," is worth 2 points, and "36" worth 1. There are seven other distracters from which the students must choose.

After 3 years, the team concluded that the advantages of this new grading system and format were substantial. First, the discrimination among answers is just as fine with partial credit machine-scoring as TA grading, and with much less variability. As proof of this, the number of regrade requests dropped from about 30 percent to zero. Second, the time to turn back exam grades is shortened dramatically, and less human effort is involved. When queried, students say their morale is higher in partial-credit multiple-choice situations, both because of the perceived increase in fairness and because they feel they are rewarded for effort.

The disadvantage, of course, is that short-paragraph responses are no longer used. However, Craig and colleagues were able to demand that students keep a course journal and use more exposition in lab reports because their TAs had more time for hand grading.

"We are well persuaded that the personal touch of hand-grading is more effective in other course activities than in exam grading," says Craig. His department now uses the time saved by machine grading to provide assistance to first-year students.

The instructors are also happy that their system discourages random guessing. With only ten choice answers and (as an example) answer choices worth 7, 5, 3, and 2, the "guess" yield is only 1.7. Finally, the instructors did not have to purchase any new equipment for their grading system. It appears that most machine-scoring systems (optical or mechanical) are readily adaptable to partial-credit scoring.

A colleague in biochemistry uses a similar method in an upper-division course.

DEMONSTRATION QUESTIONS USING LIVE DEMOS OR VIDEO CLIPS

Karen E. Eichstadt, Department of Chemistry, Ohio University, Athens, OH 45701; TEL: (614) 593-1739; FAX: (614) 593-0148;
E-MAIL: eichstadt@ohiou.edu.

Courses Taught

- Chemistry 101—"Chemistry in Today's World" (for nonscience majors)
- Chemistry 121, 122, 123—introductory course sequence for students not taking further college chemistry
- Chemistry 151, 152, 153—mainstream general chemistry

Description of Examination Innovation

On her nonscience major and introductory chemistry exams, Karen Eichstadt uses a "visual section" worth 15 percent of the exam. Demonstrations done at the beginning and end of the exam hour are followed by exam questions that require students to observe and interpret results.

Eichstadt uses demonstrations to simulate the work of a chemist within the testing situation. The exam format includes live demonstrations with close-up projection or videoclips from CD-ROM, disk, or tape. The voice-over can be muted on most media so the instructor may replay a demonstration segment for student response in observing and interpreting results.

In a typical question, students observe the behavior of certain chemical reactions and then offer an explanation for their observations based on principles discussed in the course. Eichstadt often asks them to design an experiment to test their theory and predict results from their experiment. A written description of the equipment and procedure may also be included in the question.

The instructor has a large video library with a collection of clips that illustrate relevant basic chemical principles. She also incorporates material from newspapers and journals, such as the *Journal of Chemical Education*, which links chemistry to real-world events.

Students generally perform and respond well to questions. Their most common errors include using an inappropriate principle or trying to draw too many conclusions from the data.

A set of Eichstadt's examination questions can be obtained by writing to her at Ohio University.

ABSOLUTE GRADING SCALE, UNTIMED ESSAY EXAMS

Arthur B. Ellis, Department of Chemistry, University of Wisconsin–Madison, 1101 University Avenue, Madison, WI 53706; TEL: (608) 262-0421; FAX: (608) 262-6143; E-MAIL: ellis@fozzie.chem.wisc.edu.

Courses Taught

- General Chemistry

Description of Examination Innovation

Arthur Ellis has taught general chemistry at Madison since 1977. For many years, his exams were predominantly multiple choice, with a modest fraction of the total points based on writing a few short sentences. He and his graduate TAs regarded this as a reasonable compromise that captured the spirit of the course—training in algorithms with a few insights required—and enabled them to handle efficiently large numbers of nonmajor students. He dismissed a few student complaints that the multiple-choice questions were nitpicky or inadequate measures of understanding as not representative of the class as a whole.

In 1991, Ellis taught the introductory course after having been away from it for several consecutive semesters. To his dismay, Ellis discovered that even though he had thought his teaching had not changed, there had been a substantial decline in student satisfaction based on course evaluations, with the tests a particular focus of hostility. He began to rethink the course, to ask what its objectives should be, and to read about new pedagogical approaches. At the same time, he was becoming involved in the development of new materials science content for the course and exposed to ideas of quality measurement through colleagues in the College of Engineering and Business School. Over the next few years, these influences led him to change his course substantially, including the examinations.

Ideally, the examination is a tool that helps teacher and students assess progress in attainment of the knowledge and skill objectives. In retrospect, Ellis's traditional multiple-choice-based examinations did not communicate the proper message. They were commonly viewed either as rote plug-and-chug exercises and/or, as noted by some of the students interviewed for this volume, as efforts to trick them, leading to an adversarial student–teacher relationship. Furthermore, the fact that the exams were graded on a curve, that considerable memorization was needed, and that there was a fixed time limit led to considerable test-taking anxiety for many students.

In attempting to address these problems, Ellis looked for a way to make test-taking in large general chemistry courses less stressful to encourage more creative thinking and to introduce options for guided reasoning into his examinations. The result was to grade on an absolute scale; to use an all-essay format, with essentially no memorization; to design the test so that virtually all students had enough time to finish; and to develop the Test Insurance Page (described in "Listening to Faculty") that permits students to "buy a clue" if stuck on an essay or multistep question. Clues are worth a modest fraction of the answer's value and obtained by using "scratch-off" paper such as used in instant-winner lotteries.[7]

These changes were phased in over several years and are now standard operating procedure, accepted by both students and TAs in the course. Elimination of the curve and institution of the absolute grading scale probably had the single largest effect, as they changed the atmosphere of the classroom to one of cooperation rather than competition among students: Strong students are willing to help weaker students understand course material because their assistance does not jeopardize their grade, and examinations take on a character of students competing with themselves rather than with their classmates.

The all-essay format employed in examinations has no memorization required (all formulas and constants are provided), but requires the student to demonstrate mastery of the concepts, typically by applying them to new situations. Interdependent, multipart questions are used with hints provided on the Test Insurance Page for students who are stuck. To ensure that adequate time is available, each TA takes part of the exam beforehand, and a doubling of their longest collective test-taking time, with a few incremental time additions as needed during the exam, permits virtually all students to finish in a reasonable time period.

Response to these procedures has been enthusiastic, with students now appreciating that the instructor genuinely wants them to be successful in the course. By all the metrics available to him thus far, Ellis says students are far more pleased with the course than when he began making changes in 1991.

INDIVIDUALIZED QUIZZES, CLASS EXAM RETAKES

Victor L. Garza, Chemistry Department, San Antonio College, 1300 San Pedro Avenue, San Antonio, TX 78212-4299; TEL: (210) 733-2711; FAX: (210) 733-2338; E-MAIL: v.garza@accd.edu.

Courses Taught

- General Chemistry

Description of Examination Innovation

For the past 3 years, Victor Garza has been giving his freshmen take-home quizzes (to replace graded homework) with three or more individualized an-

[7] Each student receives a "Test Insurance Page" at the beginning of an exam with clues identified by problem number, page, and point deduction. Unless "used" by the student, the clues are concealed.

swers, so that students can work together on general concepts. Each member of a group has his/her own version of the question and individualized answer. The quiz questions are an extension of mostly one-page outlines that serve as a preview, review, and lecture outline for his students.

Garza first attempted this innovation when he saw students copying answers to graded homework problems from the instructor's manual. Even when answers were not available, only a few students actually worked the problems, and the other 80 to 90 percent simply copied—sometimes as late as 10 minutes before class. With multiple versions of questions, students must plan ahead. Now, better students serve as tutors when pressed for help, showing what adjustments are needed to correctly answer different versions of a problem. Students may use any resources they want except another professor. Concepts are analyzed from different angles in the multiple versions.

Another innovation Garza has used for 10 years is permitting retests of multiple-choice examinations if the class average is very low. Or, the instructor adds questions from the earlier section on the following exam.

To protect against in-class cheating on multiple-choice exams, Garza has found it useful to hand out multiple versions of exams. Although creating the different versions is time-consuming, others at San Antonio College also use multiple versions to minimize cheating. For standardization and to measure his own performance against that of his peers nationally, he often administers an old American Chemical Society exam as a final.

Garza's colleagues applaud his efforts with these innovations. Garza is especially pleased with the added challenge and accountability he provides his more motivated students.

MODIFIED CURVE

George Gorin, Chemistry Department, Oklahoma State University, Stillwater, OK 74078; TEL: (405) 744-9845; FAX: (405) 744-6007;
E-MAIL: gochem@vm1.ucc.okstate.edu.

Courses Taught

- General Chemistry
- Physical Chemistry
- Chemical Literature

Description of Examination Innovation

"Letter grades ought to reflect a range of performance, not be based on an idealized curve that has no empirical basis," says George Gorin. "If you look at the results of almost any kind of numerical exam scores, you'll see that they not fall on the so-called 'normal' curve... because they are not derived from random processes."

So what can an instructor do? There is always "absolute" grading. But Gorin prefers his own method of adjusting raw scores, a method he used in large general chemistry classes at Oklahoma State until his retirement in 1990. Gorin makes the middle of the C range correspond to the maximum in the distribution curve, and he adjusts the range so it will comprise one-third to one-fourth of the grades. Then, he determines B and D ranges using two criteria: The number of students receiving either grade should be about half the number of C grades, and the numerical ranges should be of comparable magnitude.

In more advanced classes, which the students presumably select by interest and for which they have considerably more background, he makes the maximum correspond to the B range. "By this time, the students who are incapable or unwilling to study the material have been eliminated, so it makes sense," says Gorin, "to adjust to a higher maximum."

When asked how he would adjust for differences between one class and another, Gorin said that he would not attempt to do so. "In large public institutions, the students who populate large classes presumably represent a cross-section of the population. If the raw scores are 15 points lower this year than the previous year, that tells me something is wrong, but probably it's not the quality of the students. It could be that I didn't cover the material as well in lecture, or that the textbook we switched to doesn't discuss the material as well."

With Gorin's method, one thing is sure: Students can be certain of where they stand in terms of the class average.

Colleagues, though interested in what Gorin does, haven't switched to his method. "Everyone has his/her own way of grading," says Gorin. He's particularly sympathetic to humanities professors, who must give grades even though their subject matter doesn't always lend itself to objective evaluation, in his opinion. "I wouldn't presume to tell them how to grade," he says. "But in courses such as college science, grades ought to be based on a rational method, and not on an arbitrary standard of performance or the performance of previous students."

LINKING CONTENT TO DESIRED OUTCOMES

Thomas J. Haas, formerly of U.S. Coast Guard Academy, Vice President for Academic and Student Affairs, and Dean of the Faculty, William Penn College, 201 Trueblood Avenue, Oskaloosa, IA 52577; TEL: (515) 673-1010; FAX: (515) 673-1396; E-MAIL warde@wmpenn.edu.

Courses Taught

- General Chemistry

Description of Examination Innovation

As part of a major overhaul of the freshman general chemistry course at the U.S. Coast Guard Academy, the faculty first developed a list of "desired outcomes," including content mastery, decision-making capability, analytical problem solving, communication skills in technical writing, use of computers, "quickening of curiosity" in regard to chemistry and the other sciences, and "professionalism" appropriate to the performance of a Coast Guard cadet. According to Thomas Haas, Kevin Redig, and Robert Redig, in an article in the *Journal of Chemical Education*, these outcomes provided "a long-range vision for the program in which faculty had shared ownership" and a set of "checks and balances" against which to measure future innovations.[8]

The faculty's first major challenge was finding a way to communicate to students the faculty's expectations of them. First, the faculty developed written documentation to describe an academic routine well suited to the study of chemistry. Next, they divided the semester up into a number of exam cycles to correspond to the material covered for a given exam. At the beginning of an exam cycle, each student gets a syllabus of reading, lab and homework assignments, and a list of objectives that clearly spells out exactly what is expected for mastery.

To get feedback on students' understanding, attitudes, and goals during the semester, Haas and others also use in-class assessments such as the 1-minute paper and documented problem solutions.[9]

The results have been gratifying. Each of the several teachers instructing in the course has had to organize and direct the development of the skills necessary for mastering that portion of the syllabus. In addition, the objectives have become a yardstick by which students routinely gauge their own progress. It is no longer unusual, Haas reports, for a student to come in for extra help with a series of well-defined questions on various aspects of the material covered. The questions usually grow out of the objectives.

During exam time, all instructors have noted, students are more focused in their studying, and, even after a "tough" exam, there is no longer a chorus of complaints. Typical is the response: "It was a tough exam, but there wasn't anything on it I didn't expect." The result of the addition of objectives per section of the course is that instructors and students have become "partners" in striving for a common goal:

---

[8] "Promoting the Relevance of Chemistry: The General Chemistry Program at the United States Coast Guard Academy," *Journal of College Science Teaching* 24, no. 4 (February 1995): 249–255.

[9] Thomas A. Angelo and K. Patricia Cross, *Classroom Assessment Techniques: A Handbook for College Teachers* (San Francisco: Jossey-Bass, 1993).

"We now hold our students accountable to master a much greater level of difficulty than we have in the past" the authors write, "and we ask them, 'Do you question [a] concept on grounds of principle or because of the work it entails?'"[10]

WRITTEN QUANTITATIVE EXAM QUESTIONS

Robert M. Hanson, Department of Chemistry, St. Olaf College, Northfield, MN 55057-1098; TEL: (507) 646-3107; FAX: (507) 646-3107 (call first); E-MAIL: hansonr@stolaf.edu or http:\\www.stolaf.edu\people\hansonr.

Courses Taught

- General Chemistry

Description of Examination Innovation

Students in Bob Hanson's General Chemistry course, using the technique of "Data Driven Chemistry," write detailed quantitative essays summarizing their understanding of atoms based on a specific set of data provided by the instructor.

Students are provided with all data for the hour-long examination, which consists typically of just three questions:

1. Describe the formation of the hydrogen atom in terms of potential and kinetic energy. For full credit, be sure to be *quantitative*. That is, let's see some calculations!
2. How do ideas of probability add to your understanding of the hydrogen atom? For full credit, discuss a specific example of a hydrogen atom changing from an excited state to the ground state in terms of electron location and energy.
3. Describe how atoms containing more than one electron are similar or different from the hydrogen atom. For full credit, show *specific examples* of the phenomena you discuss.

Descriptions must be both qualitative and quantitative, and not too general (e.g., "protons attract electrons"). The instructor wants students to provide the significance of the numerical results ("Based on an energy release of 2.18×10^{-18} J, we see that if the hydrogen atom involves only potential energy, then its size is expected to be about 1.06×10^{-10} m because..."). Students demonstrate mastery by using the mathematics involved in a question to support a model and not just by plugging numbers into equations. In their essays, students must find a way to bring in the mathematics themselves.

---

[10] "Promoting...," op cit., p. 253.

Hanson uses five models to discuss atomic structure: potential energy, total energy, probability, electron spin, and energy well. Each model builds upon the previous model, helping to explain data that the previous model did not.

Students' homework consists of writing and rewriting an essay describing their model of the atom. Each day's "essay" starts again at the beginning, describing the simple model and its evolution to more and more sophisticated theory. The repetitive nature of this rewriting helps students synthesize their model and aids in retention. Students learn quickly not simply to restate the data, but, rather, to *analyze* it, synthesizing it into their model or explaining why the data might be inconsistent with their model.

Based on a modified department evaluation form, most students in Hanson's most recent class enjoyed writing essays and conducting group discussion. One student wrote: "These essays were very helpful in forcing me to learn the material well—similar to 'reteaching' what we had learned in class. It was also *great* to see actual data and analyze that to learn various theories."

Hanson says his methods are being used at Northfield High School in advanced placement (AP) and lower-level chemistry classes.

MISTAKE-FINDING EXERCISE INVOLVING WRITING, DRAWING, AND MODIFYING PICTURES; STUDENT-GENERATED QUESTIONS; 5-MINUTE LOOK-UP OPTION

Erica Harvey, Department of Chemistry, Fairmont State College, 1201 Locust Avenue, Fairmont, WV 26554; TEL: (304) 367-4498; FAX: (304) 367-4589; E-MAIL: elh@fscvax.wvnet.edu.

Courses Taught

- Introductory Chemistry
- Advanced Inorganic Chemistry
- Bioinorganic Chemistry
- Environmental Chemistry
- Physical Chemistry
- Analytical Chemistry
- Instrumental Analysis
- Introduction to Research

Description of Examination Innovation

Because Erica Harvey teaches so many different courses at different levels, in her years of teaching she has tried many different assessment techniques. In all her exams, she tries to encourage higher-order thinking and to present interesting problems that relate to real life or require pulling several concepts together. She also tends to design tests that require students to use words, sentences, and pictures to describe science concepts. Among her specific innovations are the following:

Some exam questions ask students to critique another student's work. A written-out problem solution or laboratory notebook entry is provided. Students are asked to find a specific number of mistakes or questionable assumptions and then to use these to write a summary critique. This kind of question can be take-home or in-class, depending on the level of the course. Harvey finds this a good way to reward students who have made their own mistakes (as a result of which they become more self-critical of their own work). It also teaches the risks of routine sequencing of steps to solve a problem. Finally, students tend to like this exam format and appreciate that they will need the skill of finding mistakes in their professional life.

A variation on the mistake-finding question is to ask students to critique textbook statements from lower-level courses to make them more precisely true. Examples: "Strong acids dissociate completely in solution," or "When solids and liquids are involved in a chemical reaction, their concentrations are not included in the equilibrium expression for the reaction."

Harvey also is attracted to open-ended problems that address real-life situations and require students to make and explain many assumptions using quantitative reasoning. She uses questions such as the following: "The Harvey clan was canning tomato sauce in their kitchen. They noticed that the kitchen was soon very hot. Predict the temperature rise per batch canned, and calculate how hot the kitchen was after 10 batches had been canned." Students have to answer questions such as these in three ways: (1) by writing a protocol of how to solve the problem, including assumptions and the justifications for the assumptions; (2) by providing equations that solve the problem, with all variables defined but no numbers plugged in; and (3) by solving the problem with a calculated numerical answer using appropriate values.

If she intends to ask essay questions on an exam, Harvey will distribute a list of possible essay topics from which the closed-book, in-class questions will be selected. A typical question: "Explain in words and using appropriate chemical equations how one chlorofluorocarbon molecule can destroy many molecules of stratospheric ozone." Take-home tests will require longer answers. Example: "Discuss the interplay of the first and second laws of thermodynamics for a certain global isolated system..."

Even in-class tests, for which essay questions are not previously circulated, require much writing and higher-order comparisons and contrasts. Example: "When is it advantageous to calculate lattice energies using the Kapustinskii equation, rather than the equation that is derived from the definition of lattice energy?"

When asked by students about her grading standards, Harvey replies: "You will be graded on your use of clear, concise, pertinent, logically structured, scientifically accurate, and grammatically correct prose."

Harvey also uses questions asking students to modify a picture, graph, or map. These questions play to the strengths of students who think and learn visually, and are not difficult if the student really understands the concept. An example of a drawing question: "Given this 90 percent probability contour surface, sketch a 95 percent probability surface." Or, "Fill in the missing axis label on this graph," or "Draw in the missing line on the diagram," or "Fill in the value for the missing element or species." Such questions, reports Harvey, are easy to grade.

Sometimes Harvey requests questions from students and selects some for the exam. When her class has covered a lot of detailed material from many different areas, she has given the following question: "Propose and answer a question of your own that is comparable in difficulty to the other 10-point questions on this exam."

Also available to her students are questions that require the use of primary scientific literature for take-home exams. Students are asked to discuss or critique a paper's experimental design or its conclusions, to explain a new concept from a paper, or to compare and contrast one paper's interpretation with that of the textbook.

Eager to reduce competitiveness and tension in her classes, Harvey allows students 5–10 minutes at the end of an exam period to look up any forgotten details. There is no way, she reports, that they can look up everything, but the student who's drawn a particular "blank" gets some relief. Another technique for improving students' time budgeting is that the instructor ranks the difficulty of exam questions so that students can start with the easier ones.

Finally, she employs a variant on Dudley Herschbach's "resurrection points" (see p. 100) by allowing her students to take an optional second part of their final exam, one which offers more challenging, open-ended questions. If they are within two points of a higher course grade at the end of the first (required) portion of their final, and they demonstrate substantial understanding on the optional part, they get the higher grade.

Harvey's students describe her exams as fair and fairly graded, but difficult, particularly the take-home and those questions requiring students to use scientific literature. "This isn't supposed to be an English class," students at all levels have remarked to Harvey. Harvey says that although these "utilitarians" have the most trouble with her exam style, there is a subset of students who write and think about science better than is evidenced by their grades on multiple-choice problems. These students tend to enjoy and do well on her exams.

Harvey's colleagues at Fairmont State College are interested in these various techniques, and some also use essay questions and open-ended problems on take-home exams.

"RESURRECTION" POINTS

Dudley Herschbach, Department of Chemistry, Harvard University, 12 Oxford Street, Cambridge, MA 02138; TEL: (617) 495-3218; FAX: (617) 495-4723; E-MAIL: hbach@chemistry.harvard.edu.

Courses Taught

- Chemistry 10 (upper-level introductory course)
- Chemistry 7 (second semester of introductory course for less well-prepared students)
- Chemistry 8 and 9 (special course integrating first-year chemistry and physics)

Description of Examination Innovation

In 1987, Dudley Herschbach, professor of chemistry at Harvard University and Nobel Prize Winner in chemistry in 1986, introduced what he called the "grand reform" of Chem 10, Harvard's upper-level introductory course. He describes the course changes as follows[11]: On the first day of class, Herschbach talks to his students about the nature of science. The theme of his message is that it is more important to be ardent and persistent than to be brilliant. Moreover, it is neither necessary nor desirable to be "right" at every step along the way. He tries to get his students to believe "you won't know whether science is or is not for you unless you stay around and give it and yourself a chance."

To set a different mood for the students in Chemistry 10 and to deal with the anomie and competitiveness of this typical introductory chemistry course, the instructor has initiated some other unusual innovations: He plays music (their music) as students walk into class and meets with an elected "student advisory committee" every other week for continuous feedback on how students feel about the course. In addition, he makes himself available at one of the Harvard–Radcliffe dining halls once a week for conversation with whichever students approach to talk informally about the course.

All this is meant to humanize introductory chemistry, but at least as important in Herschbach's view has been his decision to "cover less and uncover more."[12] He frequently refers to applications and focuses on the qualitative (as against the quantitative) approach to solving problems.

---

[11] Originally from a talk given by Dudley Herschbach and reported by Sheila Tobias in *They're Not Dumb, They're Different* (Tucson, AZ: Research Corporation, 1990), pp. 59–61.

[12] A phrase coined by Victor Weisskopf of MIT but now employed widely.

Herschbach also employs some literary devices to help his students understand chemistry. He has his students write poems addressing important themes or concepts such as wave-particle duality and entropy because he feels it helps introductory science students focus on originality and innovation rather than getting the "right" answer. He also uses parables to reinforce information and develop qualitative reasoning in freshman chemistry courses.[13] In his lecture on gas laws, the instructor begins with "How Aristotle and Galileo Were Stumped by the Water Pump." He tells the stories of how Aristotle, Galileo, and Torricelli explained the workings of a water pump, demonstrating how to use Toricelli's barometer and then doing measurements to explain the gas laws. Another parable, called "How Nylons Won World War II," accompanies his lecture on polymer chemistry. It tells of how the United States kept the Allied Forces from losing the war by producing synthetic rubber after being forced to do so by the Japanese conquest of Singapore in early 1942.

As regards grading, Herschbach offers two unusual gifts to his students: First, as he announces on the first day of class, students will not be graded on a curve. He wants them to compete with a standard that he, their professor, has defined (not against each other), and if they do so successfully, all of them can get As in the course. Second, and even more innovative, is his idea of "resurrection" points. Any points not earned on a particular hour exam or quiz can be "resurrected," or made up, on the final. The system works as follows. The points a student misses on a particular hour exam, the student's "unearned points," are "book-kept" on his/her individual record in such a way that the corresponding section of the final is *absolutely* increased in value by that same number of points. So, if a student scores 90/150 on the first hour exam, then the first part of the final for that student will be worth 100 + (150 – 90) = 160 points. The second part of that student's final might be worth 120 points, the third part 140 points, and so on.

Final exams, then, in Herschbach's Chemistry 10, are individualized to account for previous difficulty and to *reward* compensatory work done by students between the hour exam and the final. This is what Herschbach means when he tells the class at the beginning of the semester: "You will not lose any points on an hour exam." For students, this translates into the possibility that, however poorly they performed prior to the final, they always have the opportunity to "ace" the course.

In the first few years of the "grand reform" of Chemistry 10, several outcomes were observed.

One effect, a large increase in course enrollment, occurred even before the "new" course started in 1987. As usual, Herschbach had appeared prior to the

[13] See Dudley R. Herschbach, "Paradigms in Research and Parables in Teaching," *Journal of Chemical Education* 70(May 1993): 391–392.

beginning of the semester at a special orientation meeting for freshmen interested in science. At this meeting, he announced two of the innovations: absolute grading and the "resurrection" points. That first semester, class enrollment nearly doubled, from 170 in the previous year the class was taught, to 300. Herschbach and his colleagues attributed this increase directly to the "news" about grading. Previously, many students who had qualified for Chemistry 10 (on the basis of Advanced Placement scores or their scores on a qualifying exam) had opted, instead, for the lower level introductory course, put off by Chemistry 10's reputation for being "cutthroat."

Despite the near doubling in enrollment, the students in Chemistry 10 have, on average, performed better than their predecessors. This has been documented both by their (absolute) final grades and their TAs' impressions. (A number of chemistry TAs teach 2 years in a row at Harvard.) Morale has been noticeably better and, as for persistence in the major, Herschbach reports that enrollment in chemistry at the junior level (physical chemistry, the first course in the standard sequence for majors) has reversed its previous decline and is on the rise. Although Herschbach doesn't think Chemistry 10 is the major reason for the change, he says, "At least it does not seem to have done much harm."

"HELP FOR A FEE" FOR NUMERICAL PROBLEMS

Vickie L. Hess, Department of Chemistry, Indiana Wesleyan University, 4201 South Washington, Marion, IN 46953-9980; TEL: (317) 677-2301; FAX: (317) 677-2284; E-MAIL: vlh@barnabas.indwes.edu.

Courses Taught

- General Chemistry

Description of Examination Innovation

Vickie Hess thinks chemistry exam problems should be true problems, not mere exercises. At the same time, she believes that students who struggle with problem-solving skills but have a mastery of the basic relationships of chemistry should not fail. So, she has instituted a "help for a fee" system on the numerical problem sections of her exams.

During the exams, she remains in her office. Students may come in with questions on problems. She does not actually deduct points for the information they come up with, but she "charges" points, varying with the degree of help, when she has to provide basic formulas and/or overall direction to help them solve an exam problem. Students who are ill prepared do not benefit much from this assistance because there isn't enough time to relearn a segment of the course in Hess's office. However, the student who is stumped by the wording of a problem, needs help in putting the pieces together, or has forgotten a specific equation, can salvage some success.

Often students are helped, Hess finds, simply by having someone to talk to. And those who are suffering exam anxiety benefit from having a break in the exam environment and a sympathetic ear. She reports less "bargaining" now than she had earlier with this approach. There was a time when some of her stronger and more assertive students would ask her, "What will it cost me if you tell me...? But suppose you only tell me...?" Hess doesn't know why this change has occurred.

Because of an increase over the years in class size, more students are using Hess's option during an exam period. So far, she has been able to handle the numbers by limiting question length, checking on the size of the line as students come in, and giving preference to students who have not yet been in. She also gives a little information and encourages student to have a go at it and come back if necessary, rather than hovering. She feels that these changes to accommodate slightly larger classes, have actually made the system more effective. "It was never intended to be an enabler for the unprepared," she says.

If her classes became very large, and if she had TAs, Hess thinks, given the resemblance among student questions, that she could go over the exam in advance with the TAs and determine most of the "fees" fairly systematically in advance.

Hess has been using this "help for a fee" technique for the past 10 years and believes it has helped "humanize" what is viewed by students as a difficult course. Students have told her they appreciate the help.

PERIODIC TABLE "CRIB" SHEET, STUDENT ADVISORY COUNCIL

Tom Holme, Department of Chemistry, University of Wisconsin–Milwaukee, P.O. Box 413, Milwaukee, WI 53201; TEL: (414) 229-3970; FAX: (414) 229-5530; E-MAIL: tholme@alchemy.chem.uwm.edu.

Courses Taught

- General Chemistry
- History of Chemistry
- Physical Chemistry

Description of Examination Innovation

Because students tend to be concerned about remembering equations, Tom Holme allows them to bring to the exam a periodic table on which they may write anything they want. However, they must write in nonerasable ink on the periodic table and use the same one (made of cardboard stock paper) for all four exams and the cumulative final. This forces students to budget their space on the card, encouraging efficient study habits.

Students who misplace a table may use a "fresh" one for data purposes only. The "one and only" table policy—borrowed from an original idea by Ron Batstone Cunningham at the University of South Dakota—is clearly stated in the course syllabus and on the first day of classes. In 5 years, fewer than 10 students out of nearly 1000 have lost their periodic table.

This practice alleviates students' concerns about memorizing equations and has been appreciated by students. Holme has noticed that once students get used to the approach, they do a *much* better job of organizing their material. Periodic tables from the second semester of the course validate Holme's premise: The notes are more sophisticated. Still, some students do not practice the rules they write down on their tables. For example, rules for chemical nomenclature routinely show up on periodic tables, but Holme still sees plenty of errors in nomenclature.

Based on an idea suggested to him by Paul Holmgren, biologist at Northern Arizona University,[14] Holme also employs a student "advisory council" to handle complaints about test items. Six to eight students from a large lecture class participate on the council. Membership is determined by student nomination, statements from nominees, and the instructor. At test time, the instructor has the advisory council approve the answer key before he grades the exams. During the exam, each student is given an extra piece of colored scratch paper for concerns about a question. These comments are put in a box when the test is handed in and go directly to the council. Holme *never* sees them. The advisory council then takes the proposed key and all the student comments (usually about 10 percent or fewer write comments), and addresses the issues raised.

In order for Holme to accommodate a student's concern, the council must provide documented evidence as to the unfair nature of the question (and "hard" does not equal "unfair," says Holme). If they succeed, the instructor bargains for a remedy: two possible answers to a question, entire credit, or partial-credit points for essay questions.

Holme has found the council to be one of the most effective tools he has for gauging how a class is doing and when frustration is causing a learning gap. And he is virtually never accused of being unfair.

---

[14] Described in Paul Holmgren, *Journal of College Science Teaching* 21(1992): 193.

TWO-TIER SYSTEM, TEAM QUIZZES, HELP NETWORK, TAKE-HOME FINAL EXAM SECTION

Kimberly L. Kostka, Department of Chemistry and Geology, University of Wisconsin–Rock County, 2909 Kellogg Avenue, Janesville, WI 53546; TEL: (608) 758-6532; FAX: (608) 758-6560; E-MAIL: kkostka@uwcmail.uwc.edu.

Courses Taught

- General Chemistry
- Humanities Chemistry

Description of Examination Innovation

Kimberly Kostka uses what she calls "a two-tiered exam system" in her general chemistry classes. In the first tier, students are tested *without numbers*: In-class, timed exams include only open-ended questions. The students also take a "second-tier" exam, which is a cooperative, weekend take-home exam covering the same material. About half of Kostka's exams are in each tier, usually alternating exam by exam.

In the first-tier exam, open-ended questions typically ask for definitions and statements about chemical concepts. The students are expected to support their statements with examples that demonstrate mastery of the material. Also on these exams are application items requiring students to apply concepts to solve problems—again, *nonnumerically*. As an example, Kostka says that after discussing buffers, she will include on an exam a question asking students to outline a research plan that would allow a chemist to determine the preparation, pH range, and toxicity of a buffered intravenous drug. Students must also explain the research and predict the results. These questions, says Kostka, "require the students to think critically and creatively about the problem while they use the concepts they are learning."

The second-tier take-home exam requires mathematical analysis of conceptual problems. Kostka says, "I encourage my students to work together and warn them against working unfairly," by which she means using other people's information without giving credit. She allows them to use any printed resources they wish, including the textbook and their notes, as long as they cite their resources, printed or spoken.

"I give the exam this way because it allows students ample time to carefully consider the results of their calculations. In fact, I specifically ask them to comment on the 'sense' of their answers," she says. Her exams typically include more challenging problems than Kostka would normally put on an in-class exam. But she says students, especially those who are nervous about their mathematics, still prefer this format because of the extra time.

"The take-home exam provides them with time to learn *while they are being tested*," Kostka explains. Indeed, they are freed from many of the pressures associated with timed exams. Kostka seems particularly interested in improving the culture of science courses so that there is a "sense of community and responsibility" in her classes. Collaboration, she feels, is essential to that end. By encouraging it, "students become 'plugged in' to their chemistry community. This personalizes their chemistry education."

Kostka has also tried to use team quizzes, but the overwhelming student response was negative, she reports, *despite an increase in the average class grade*. In fact, the more able students who were paired with weaker students chose to take a grade penalty rather than pull the weaker student along. One of her strongest students willingly allowed his unprepared partner (as well as himself) to fail as a lesson.

This surprised the instructor. Kostka then modified her original approach, which had randomly paired students doing in-class quizzes together. Now, students are given a quiz to work on independently. After 10 minutes on their own, they are allowed to work with anyone else they choose for the remaining 5–10 minutes. Kostka does not always use this format, however, and she does not announce of ahead of time when she will allow collaboration.

"This method ensures that each student will work hard to identify his/her own strengths and weaknesses *before* teaming up with others," she says. "They must make their own intellectual investment before working with others."

Because she teaches only one lecture section of approximately 45 students, Kostka hasn't been able to do a cross-class comparison to other sections of general chemistry. However, differences between the first and second semesters are suggestive: a 7 percent higher average grade, increased self-criticism, greater use of outside help, and a stronger sense of community. Although these observations may simply reflect students' customary adjustment to college life over the first year and the dropout of poorer students between the first and second semesters, Kostka is convinced that the different testing format encourages a greater sense of community in the second semester.

GROUP QUIZZES WITH HIGHER-ORDER THINKING PROBLEMS

Sandra L. Laursen, Department of Chemistry, Kalamazoo College, Kalamazoo, MI 49006; TEL: (616) 337-7020; FAX: (616) 337-7251; E-MAIL: Laursen@kzoo.edu.

Courses Taught

- Introductory Chemistry
- Physical Chemistry

Description of Examination Innovation

When Sandra Laursen began using group quizzes in her introductory chemistry course, she was able to give students more problems that involve higher-order thinking than before. Indeed, rather than having her students perform calculations or define terms, she now is asking them to draw conclusions from data (using calculations or defining terms in the process), to find flaws in or assess scientific arguments, and to design their own experiments to address scientific questions.

A 1-hour group quiz typically consists of a straightforward exercise requiring factual information or familiar computations to "warm up" students for one or two "higher-order thinking" questions such as the following:

> Some experimental observations comparing different atoms are reported below. Write electron configurations for the atoms, and use these configurations to explain or make sense of the observations. Use spectroscopic or orbital box notation as appropriate for the question.
>
> a. Carbon and silicon have similar chemical reactivity—for example, both tend to form four covalent bonds.
> b. Both chromium and vanadium are paramagnetic, but chromium is more strongly paramagnetic.
> c. The first ionization energy of Na is 419 kJ/mol; the first ionization energy of Ne is 2081 kJ/mol.

Group assessment grew out of Laursen's increasing use of cooperative learning exercises in and out of class, such as in-class problem solving in pairs, lab partnerships, and cooperative research projects designed to help students learn to collaborate. Laursen says, "If as an instructor I think collaboration is a valuable skill and ability, I must award credit for developing and applying that skill." She assigns students to semipermanent, mixed-ability groups of four or five in the beginning of the term. On a quiz day, all students receive a copy of the quiz, and the groups disperse to work on it together. They return to the classroom at the end of the hour and hand in a *single copy* of the quiz signed by the group members. All participants in a group receive the same grade. Since the group quizzes are short, they are not particularly time-consuming to grade.

Laursen says the opportunity to discuss and refine answers tends to draw more thoughtful and well-developed responses from students and probably brings the majority of them to a higher level than would be achieved otherwise. New information can be presented and questions designed to lead students to look meaningfully at new applications or aspects of the material.

She uses the group quizzes, then, just as much as a means of teaching as of evaluating students. To prepare her students for group work and group assessment, Laursen makes a point of discussing during class the behaviors and attitudes that are helpful in working in groups.

According to course evaluations, students perceive their performance to be as good or better on group quizzes (and it *is*, Laursen says), believe they learn from them, and enjoy them more.

Laursen's desire to make students responsible for their own and others' learning, to work with others, and to communicate their scientific ideas, is reflected in the fact that group quizzes comprise up to 22 percent of her final course grades. However, Laursen alternates group quizzes with standard, in-class individual quizzes, so student grades aren't dependent solely on group work.

Some of Laursen's colleagues—principally those who also use cooperative learning—have responded favorably to her methods. Others remain skeptical.

HANDS-ON LAB QUESTION ON FINAL EXAM, "TEACH BACK"

Marya Lieberman and Seth Brown, both formerly of University of Washington, Department of Biochemistry, University of Notre Dame, Notre Dame, IN 46556; TEL: (219) 631-4665 (Lieberman), (219) 631-4659 (Brown); FAX: (219) 631-6552;
E-MAIL: mlieber@darwin.helios.nd.edu *or* seth.n.brown.114@nd.edu.

Courses Taught

- Chemistry courses at MIT and University of Washington

Description of Examination Innovation

When Marya Lieberman and Seth Brown taught as TAs at the University of Washington in Seattle, they included the following lab question to emphasize the link between microscopic events (chemical mechanisms) and macroscopic results.

> You will get a cup labeled PVA that contains 25 ml of a solution of polyvinyl alcohol in water. Polyvinyl alcohol is a polymer whose structure is illustrated below. You will also receive a cup labeled B that contains 5 ml of 0.105 M borax solution. Borax dissolves in water to form $B(OH)_3$ and $B(OH)_4$, whose structures are also shown below:

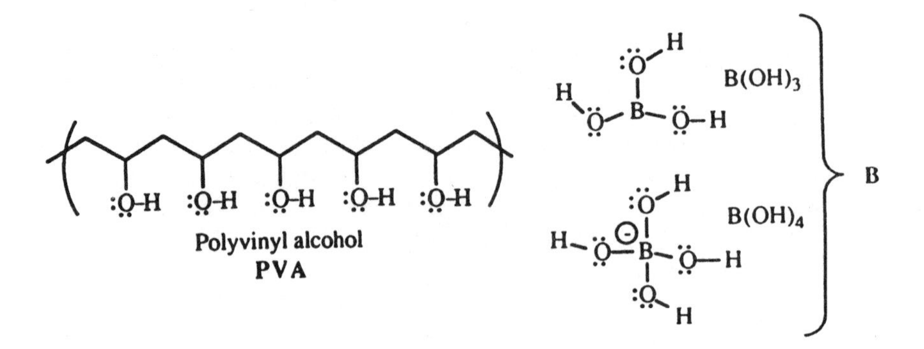

While stirring the polyvinyl alcohol solution, pour the borax solution into it. Continue stirring for a few minutes.

a. (15 pts.) Briefly describe the changes in physical properties that occur.

b. (25 pts.) Propose a chemical explanation for these changes. Remember, a picture can be worth a thousand words!

The outcome of the experiment in this particular case was "slime," a slippery polymer that the students had not previously encountered. The ingredients pose no safety hazard, although, with 25 students, one spill had to be mopped up. The advantage of asking a question such as this one, the instructors believe, is that the problem requires students to use their powers of observation and to propose an explanation based on their new knowledge of chemical principles (e.g., the reactivity of borates, the nucleophilic nature of hydroxyl groups, and the macroscopic consequences of cross-linking molecules). "We could have just described the results and eliminated the messy part, but the students clearly enjoyed this problem as a break from the paper-and-pencil portion of the exam," say the instructors.

The disadvantages were the potential for mess and space requirements for chemical storage until students are ready to do the problem. Also, it may not be possible to use such questions in large classes.

In addition, the instructors included several projects counting for about as many grade points as the final exam. "Projects allow students to put sustained effort into a substantial problem," they explain, "which requires a more sophisticated command of the subject than answering questions on an hour exam."

When Lieberman taught students at MIT, her favorite method of evaluating their knowledge was one she learned from her tutors, called a "teach back." In place of a midterm or final, a student would prepare a lecture on some topic in first-year chemistry, and Lieberman would play "student" for an hour or two. She'd ask students how their presentation connected to other concepts from the course, or for real-life examples.

The teach back differs from an oral exam in that it is the student's responsibility to structure the discussion with the aim of teaching (the teacher) a chemistry concept. Three advantages result: (1) The instructor gets a sure sense of how well the student understands the concept; (2) there are enough connections among concepts that the test rapidly becomes fairly comprehensive; and (3) because it is one-on-one, it allows the instructor greater flexibility in administering and grading the exam. (Lieberman used this technique with equal success with a blind student and with students of limited English proficiency.)

Unfortunately, teach backs cannot be administered in large classes because of the great deal of time required both to put the student at ease and then to listen to all the students' "lectures." Grading also becomes more subjective.

CONCEPTUAL QUESTIONS

Mary B. Nakhleh, Department of Chemistry, 1393 Brown Building, Purdue University, West Lafayette, IN 47907-1393; TEL: (317) 494-5314; FAX: (317) 494-0239; E-MAIL: mnakhleh@vm.cc.purdue.edu.

Courses Taught

- General Chemistry I and II (for majors)

Description of Examination Innovation

Mary Nakhleh's testing innovation is part of project REMODEL, in which she works with the other professors who teach the two-semester-long general chemistry course. The project is currently in its sixth semester. Nakhleh and her colleagues implemented and evaluated innovations in three areas: lecture, laboratory, and assessment. Of interest in this description are the interdependent innovations in examinations and lecture.

The examinations in these courses have always been composed of free-response questions, but the content of the exams was changed from highly mathematical, largely algorithmic problems to a mix of algorithmic and conceptual problems. In the conceptual problems, students are sometimes required to use words and diagrams to explain processes on a molecular level, such as drawing, on a molecular level, the species present in an aqueous solution of a weak acid. In other problems, students might be asked to use their understanding of science concepts to predict or explain phenomena, such as an explanation of the strengths and weaknesses of the Bohr model of the atom.

This shift from algorithmic skill to conceptual understanding in the examinations required that Nakhleh and her colleagues also change the lecture component of the course to give students practice in using their understanding of chemical concepts—what one professor termed "chemical reasoning." Therefore, the instructors instituted weekly special sections in lecture, in which students engaged in interactive group work to increase their understanding of the lecture material. There were five to six students per group and three sets of problems per session, so several groups could be working on one problem. As an example, one problem required that students come up with a set of desirable properties for antifreeze and then to select an appropriate chemical from a list of five substances. Groups presented their answers to the class for discussion.

For the past three semesters, students have indicated on evaluations that the conceptual questions are worthwhile and that the interactive sessions help them learn how to use chemical reasoning in answering this type of question. The students reported that they had not previously learned chemistry in this way, and it took them a few weeks to learn how to answer these more open-ended questions. The professors in the course were initially worried about time to cover the material, but all of them have reported at the end of the course that they were able to cover 95 percent of what they had planned. The professors have also enjoyed interacting with the students in the special sessions, and these special sessions seem to have helped break down the "professor-student barrier" that sometimes prevents communication in science courses.

PICTORIAL AND DIAGRAMMING QUESTIONS, ABSOLUTE GRADING, "RESURRECTION" POINTS, GROUP AND TAKE-HOME EXAMS

Susan C. Nurrenbern, Department of Chemistry, Purdue University, 1393 Brown Building, West Lafayette, IN 47907; TEL: (317) 494-0823; FAX: (317) 494-0239; E-MAIL: nurrenbern@chem.purdue.edu.

Courses Taught

- General Chemistry
- Instrumental Methods of Analysis
- Curriculum and instruction courses for chemistry workshops for middle-school teachers

Description of Examination Innovation

Susan Nurrenbern, whose early assessments of in-class examinations partly inspired this volume,[15] began to think about and develop "nontraditional" and nonalgorithmic exam questions while a visiting assistant professor at the University of Missouri–Kansas City in 1979–1980. At that time, she began to explore ways of testing students' understanding of macroscopic phenomena at the microscopic level. This fit what had been revealed in her research for her doctoral dissertation on problem solving. She maintained her interest in this area throughout her 13 years as a chemistry faculty member at the University of Wisconsin–Stout.

For general chemistry courses, the instructor allowed students to bring to exams a 3" × 5" notecard prepared in advance with information they thought

[15] See S. C. Nurrenbern and M. Pickering "Concept Learning vs. Problem Solving: Is There a Difference?" *Journal of Chemical Education* 64(1987): 508–510; "Designing Exams to Encourage Reasoning Skills Development, paper presented at the Symposium on Creative Examinations, 191st National American Chemical Society Meeting, New York, April 1986; S. C. Nurrenbern, *Experiences in Cooperative Learning: A Collection for Chemistry Teachers* (Madison, WI: Institute for Chemical Education, 1995); S. C. Nurrenbern, D. Frank, and C. Schrader, *Test Bank for Health Chemistry* (a high-school textbook) (Lexington, MA: D. C. Heath, 1993).

would be useful for the test. This greatly relieved stress, according to the instructor, and had the added advantage of being a useful study and review strategy for students.

Then and now, Nurrenbern designed questions involving molecular-level representations of macroscopic properties. Some were in the multiple-choice format. Some required students to complete a diagram. A typical question asks students, "Which diagram represents a mixture that is 75 percent Ne atoms?"

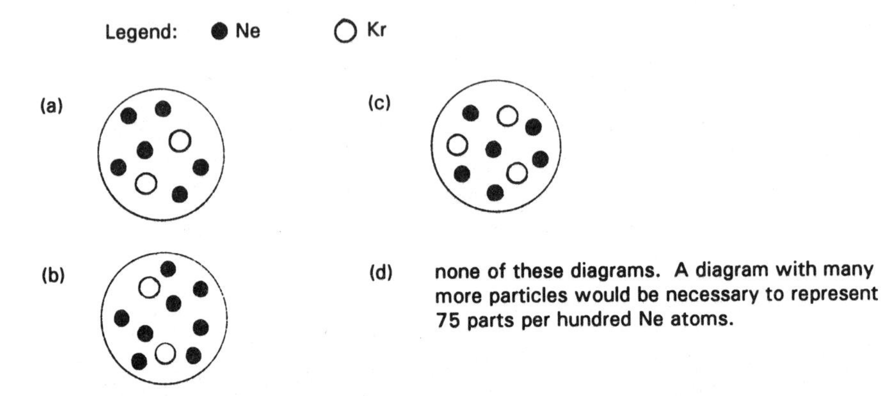

While using these "nontraditional" testing techniques, Nurrenbern also designed exams and assigned credit to problems and questions on each exam so the proportion of memorization and algorithmic questions to understanding questions was such that a student could earn a C, or possibly a B, by memorization and application of algorithms alone. To earn an A, students had to correctly answer understanding questions in addition to the algorithmic questions.

In addition, the instructor implemented an absolute grading scale, which she prefers to curved grading for several reasons: First, absolute grading reinforces the necessity that she design reasonable exams so that students can achieve at a desired level; second, it establishes an environment in which everyone can do well; third, it gives students a firmer sense of their standing and their potential "highest" grade; and finally, it makes students feel more in control of their success, not subject to some anomalous performance of their classmates.

The instructor also awarded "resurrection" points (a method devised by Dudley Herschbach—see p. 100) in general chemistry courses at the University of Wisconsin–Stout to accommodate the different rates at which people learn chemistry. Because, with this practice, students have an opportunity to make up points during the final, Nurrenbern observed an increase over the semester in student motivation with this technique.

Nurrenbern's interest in alternative testing broadened as she explored cooperative learning to address the needs and styles of nonscience students in chemistry at University of Wisconsin–Stout, where there was no chemistry major. After employing cooperative learning in general chemistry laboratories with positive outcomes, Nurrenbern developed an evaluation strategy for laboratory learning based on an open-lab notebook, take-home, cooperative laboratory midterm and final. Students predict experimental results, interpret data, and draw conclusions based on information given in the exam. One question on the exam asks each group member to write a separate evaluation of his/her experience in the group, including a self-evaluation of his/her contribution to the group.

Later, in working with middle-school teachers at the Institute for Chemical Education workshops, Nurrenbern used cooperative learning and cooperative assessment to lessen the stress experienced by teachers during "testing." Group presentations and group construction of concept maps were used to assess learning beyond memory and recall, and to reduce test anxiety.

This year at Purdue, Nurrenbern uses an "Answer Basket" for mathematical questions on multiple-choice exams taken by 1200 general chemistry students at one sitting. An "Answer Basket" is constructed for numerical problems by arranging all correct answers, and some incorrect numbers, in ascending order in a 5 × 5 matrix, with the rows labeled (a), (b), (c), (d), and (e). The letter corresponding to the row in which the answer is found is recorded on a computer scan sheet. Thus, the same number can be the correct answer for more than one problem.

"While we know it would be better to have students show their work for a problem and to grade the students' work, the challenge of maintaining reliability and validity while grading over a thousand papers is discouraging," says Nurrenbern. The "Answer Basket" represents a compromise between hand grading and the negative aspects of standard multiple-choice questions. It minimizes "testwise" students' ability to guess from among four to five alternatives by forcing them to work through the entire problem. Nurrenbern's students have confirmed this effect, and some do not like the method for that reason.

TAKE-HOME CHEMICAL SAFETY EXAMS

Maria Pacheco, Anne Marie Sokol, and Michael Cichon, Department of Chemistry, Buffalo State College, 1300 Elmwood Avenue, Buffalo, NY 14222; TEL: (716) 878-5204/5101/5113; FAX: (716) 878-4028; E-MAIL: pachecmd@snybufaa.cs.snybuf.edu.

Courses Taught

- Analytical Chemistry
- General Chemistry
- Physical Chemistry

- Chemistry (for nonscience majors)
- Laboratory Techniques in Secondary Science Education (with Anne Marie Sokol)

Description of Examination Innovation

At Buffalo State College, specific chemical safety instruction was not provided as a separate course or formally presented in any of the chemistry courses. For this reason, in 1993–94, a safety component was introduced into the general chemistry laboratory (and in analytical chemistry for sophomores and juniors in spring 1995). The course provides basic safety information in a generalized multidisciplinary environment.

The instructors in chemical safety do not expect their students to know everything there is to know about hazardous materials, but want them to know how to obtain the information they need. With this overall objective in mind, five take-home exams (with approximately 1 week to complete each) are given during the semester and are factored into students' final grades to encourage student participation and to emphasize the importance of the material. Students work together on exams and use the text available on reserve in the library and in the chemistry stockroom to assist them as well.

The first exam consists of specific questions dealing with existing laws for protection of employees, classification of materials and hazardous chemicals, entry routes, and exposure classification and prevention. The second exam introduces the Material Safety Data Sheet (MSDS) to the students. Students are asked to present and explain all the sections and information available in an MSDS. The third exam introduces the "fire diamond" and its uses.

The fourth exam provides students with hands-on experience using an MSDS. Students are asked to identify a substance; name all components and hazards; present the physical, fire, and explosion data; and indicate examples of chemical incompatibilities and ingestion hazards. They also have to identify one spill, leak, or disposal procedure for the substance. The fifth exam presents a hypothetical situation in a general chemistry laboratory. Students are asked to identify and propose ways to correct safety violations presented in a hypothetical scenario by identifying administrative, engineering, and personal protective equipment problems.

Grading is done by the instructors, not by TAs. Students are required to cite the book page for each answer to ensure that the book is being read. After corrected exams are returned, students have the opportunity to make corrections and resubmit the exam for regrading. The instructors' goal is for students to prove they were able to find the information, and leave with five corrected exams in hand to use as future reference.

The safety tests are currently given to all the new TAs in the chemistry department to assist them in preparing their prelab safety lectures. It also enhances their knowledge, important for international teaching assistants who may not be familiar with U.S. chemical safety regulations. The inclusion of safety information for analytical chemistry in spring 1995 was done in conjunction with the Environmental Health and Safety officer on campus. The officer was impressed with the knowledge retained by the students as a result of the testing scheme.

GROUP TESTING

Mark A. Ratner, Department of Chemistry, Northwestern University, 2145 Sheridan Road, Evanston, IL 60208-3113; TEL: (708) 491-5652; FAX: (708) 491-7713.

Courses Taught

- Physical Chemistry
- General Chemistry (for first-year students)

Description of Examination Innovation

While most of Mark Ratner's testing is rather standard, he finds one innovation, which he has used from time to time in relatively small classes (no more than 25 students), extremely valuable: the study group/class fragmentation approach that originated in the Berkeley Mathematics Department.

In each study group of four or five people, the instructor asks one person to take primary responsibility for a particularly knotty concept in physical chemistry. All study groups receive the same set of topics, which range from fairly general (definition of the chemical potential, temperature as an index for equilibrium among energy levels) to more specific (zero-point energies and quantum effects on ground states, the nature of black-body radiation as an indication of the failure of classical mechanics). The student "expositor" for a particular concept is responsible for teaching his/her study group, in one or more sessions of 1–2 hours outside of class, about this concept. Students are required to do the appropriate text reading beforehand, and the other members of the group are supposed to ask questions and explore applications. The sessions take place either in the instructor's office or in a small conference room, usually in the evening, with the class instructor present at (minimally) one (preferably the first) of these teaching sessions.

Class examinations then include questions on these topics, as suggested by the subgroup of student "instructors." (The instructor "massages" their questions, he admits, but he does ask the students who created them to assist him in grading.) Students estimate spending between 4 hours to a full week on this assignment; Ratner believes a realistic time period for most topics is about 10 hours. He has not encountered any resistance from students, though the instructor has learned to give them ample warning early in the term about the assignment.

Several pedagogical goals are advanced by this technique. But most important, Ratner insists, is that it encourages students to take proprietary interest in the learning accomplished by the class. "When students are responsible for teaching part of the material and see some of the difficulties entailed in exposition, they identify more closely with the goals of the course and their interest level is substantially higher than usual," he says.

RETESTS FOR WEEKLY QUIZZES

Maureen Scharberg, Department of Chemistry, San Jose State University, San Jose, CA 95192-0101; TEL: (408) 924-4966; FAX: (408) 924-4945; E-MAIL: scharbrg@sjsuvm1.sjsu.edu.

Courses Taught

- Chemistry for Nonscience Majors

Description of Examination Innovation

Maureen Scharberg gives 30-minute multiple-choice "unit" quizzes in her chemistry classes every week, with a retest option for scores below 80 percent correct. These quizzes, which are based on lecture and lab material, are helping nonscience majors learn basic chemistry better. And surprisingly, many of the students say they like getting quizzed this often because it forces them to study the material each week rather than all at once, right before a major test.

In the beginning, Scharberg used questions from a test bank devised by her predecessor for this purpose. Now, however, she says she's gotten "quite good" at devising questions for the quizzes. All the problems are straightforward. For example, students are told that a particular washing soda has a pH between 8 and 10, and students must choose the correct classification for the washing soda from a range of five options (strongly basic to strongly acidic). In another item, students must select an accurate description of a colloid such as whipped cream from five choices (a gas in a liquid, a gas in a solid, etc.). Twenty problems such as this are on each quiz.

The quizzes are given during weekly laboratory periods so that Scharberg or one of the lab instructors can give individual help to students, something that could never be done in a large lecture with between 250 and 320 students possessing a wide range of interests and abilities in science. The quizzes are computer graded and account for a significant 38 percent of the students' final grade.

Most students take advantage of the retesting option, which Scharberg thinks can get them to learn material they would otherwise not return to until the final exam. Some students, however, are satisfied with a B. In her lab section last term, 13 out of 24 students were eligible for a retest, and only 9 did so.

"It seems to depend on their schedules. And some students consistently retest, while others consistently do not," says Scharberg. She feels retesting helps nonscience majors get over the obstacles they may encounter in science classes.

"With chemistry, nonscience students are confronting (some for the first time) their fear of chemistry, their fear of math, *and* their fear of taking tests. The retesting option compensates for that a little bit."

Retests are "equivalent" quizzes, with similar questions—same general concepts, same level of difficulty. These quizzes don't take time away from laboratory, Scharberg notes, because students go over the quiz in small groups when they aren't busy, such as during cleanup, so laboratory work is not compromised.

Scharberg has been using this system for 4 years and reports that it has been fairly successful. However, she noticed that some students were "relying too heavily" on the retesting option. They used the first quiz as a sort of "dry run" and did not study for it at all. Then they fell behind: They weren't able to "catch up" for the retest. Others didn't do any better on their retests than they did the first time around, although they went over their errors with the lab instructor, indicating to Scharberg that they simply weren't studying outside of class.

To correct this, she now runs a small group discussion for students who don't score well on a unit quiz. She gives these students individual help and extra problems before they take the retest. "It's a sort of group tutoring session, and I think it's helping," says Scharberg.

Students who do worse on a retest are not penalized, though Scharberg says this doesn't happen often. Overall, students are improving their scores and seem to be learning how chemistry plays an important role in their everyday lives.

GROUP QUIZ

Leslie Schwartz, Chemistry Department, St. John Fisher College, 3690 East Avenue, Rochester, NY 14618; TEL: (716) 385-8237; FAX: (716) 385-7311; E-MAIL: schwartz@sjfc.edu.

Courses Taught

- General Chemistry

- Physical Chemistry
- Quantum Chemistry
- Nursing Chemistry
- Modern Physics
- Quantum Mechanics and Reality (interdisciplinary course for non-science majors)

Description of Examination Innovation

In the fall of 1993, Leslie Schwartz started giving her freshmen in general chemistry a group quiz every class period on the textbook reading and homework problems that were due for the previous class. The quiz questions are not difficult, but they generate heated group discussion and debate—which is just what Schwartz is after: "You see lively, serious discussion of questions you've specifically designed to accomplish some learning goal."

Each group, consisting of three to four students, hands in one quiz, and all students in a group get the same grade. Students' satisfaction with the groups themselves predicts satisfaction with the group grade, says Schwartz. She tried both random and self-selected groups, and her students overwhelmingly preferred the latter.

The fact that these quizzes count for 40 percent of the final grade motivates students to keep up with the work, though Schwartz graciously drops the two lowest quiz grades before determining final course grades. Schwartz also gives three individual exams, one final exam, and daily homework sets. Because she feels the homework problems are "a learning, more than an assessment, tool," Schwartz grades these only on effort.

Questions on the group quizzes are both calculational and essay types. Those that work best in the group setting ask for an explanation of a calculated answer or phenomenon, such as the following question:

> Which solution is the most concentrated? Explain.
>
> i. 6.0 mL of 2.0 M ethanol
> ii. 4.0 mL of 4.0 M ethanol
> iii. 3.0 mL of 5.0 M ethanol.

Schwartz used to give individual daily quizzes, so she feels she's saving a lot of time using group quizzes. Also, by the end of the course, the quizzes become such an important part of the learning process that they take up at least half of each class period, and grading is more manageable.

The majority of Schwartz's general chemistry students like that the quizzes were both daily and cooperative. In end-of-course evaluations, students remark that the quizzes keep them thinking about the material and give them a good idea of how well they understand it. Some admit they rely on others too heavily, and a number of students suggest adding individual quizzes on a regular basis to keep everyone on track, which Schwartz intends to try next.

Schwartz got the idea for daily quizzes from a colleague who gives them in both freshman chemistry and organic chemistry. His quizzes are shorter and individual, as he fears group quizzes can lead to false confidence on the part of weaker students. Schwartz values the collaborative learning component enough to keep working at how to use it best and encourages cooperative learning in her upper-division physical chemistry classes as well.

REQUIRED RESEARCH PAPER, ORAL PRESENTATIONS, MULTIPLE ASSESSMENT MODES

Shirish Shah, Department of Chemistry, College of Notre Dame of Maryland, 4701 North Charles Street, Baltimore, MD 21210; TEL: (410) 532-5712; FAX: (410) 435-5937.

Courses Taught

- Chemistry and the World Around Us (nonscience majors)
- Chemistry for Allied Health and Nursing Students
- General Chemistry
- Organic Chemistry
- Environmental Science (nonscience majors)
- Technical Physics
- Physical Chemistry
- Analytical Chemistry
- Chemical Principles I and II (nonscience majors)

Description of Examination Innovation

For the past 3 years, Shirish Shah has been requiring a research paper and its oral presentation in all of his chemistry classes. Topics are selected in cooperation with the instructor. In his "Chemistry and the World around Us" course for nonscience majors, he requires a short project on how government agencies, legislative bodies, private industry, and nonprofit organizations deal with environmental problems. He and his students view this oral presentation as the capstone of the course, demonstrating why they are learning chemistry and how it affects daily life.

In the same course, he also requires a group library project for which he selects and puts on reserve journal articles on current topics. Groups of three to four students choose a single topic such as science and women, global warming, ozone depletion, rain forests, or "comparative" educational systems. To ensure some degree of individual accountability, each student must write a 2-page paper summary of the presentation and recommendations, and a lead student must coordinate each group presentation.

At the beginning of the semester, Shah tends to give at least a couple of take-home tests (of between 60 and 70 questions). These questions require short essay answers based on library research, in addition to referencing the course text. This process assists students in the selection of a research project topic, as well as a subject for the group project.

Grades in Shah's courses are absolute. Due dates for projects, quizzes, and exams, plus percentages per project and dividing lines for A's, B's, C's, and D's are explained early. Students are provided with a syllabus, complete notes from the instructor (allowing students to focus on listening and participating), and lab procedures during the first week of classes. They know what is expected of them and when, lessening the anxiety they may feel over grades.

Since there are several different assessment modes—quizzes, tests, projects—Shah finds that students' "math anxiety" has been reduced. Completion rates in his courses are now 95 percent, compared to under 80 percent prior to 1991–1992. And as retention increases, so does enrollment—up from 85 in 1991, to 185 in 1995.

LAB AND ORAL REPORTS ON EXPERIMENTS, "ACTION" TESTS, STUDENT-GENERATED TEST QUESTIONS

Sharon L. Tebben, Assistant Dean of Education, Department of Elementary and Secondary Education, Northern State University, 1200 South Jay Street, Aberdeen, SD 57401; TEL: (605) 626-7685; FAX: (605) 225-2741;
E-MAIL: tebbens@wolf.northern.edu.

Courses Taught

- Introductory General Chemistry
- Inorganic Chemistry
- Physical Science for Elementary Teachers

Description of Examination Innovation

Sharon Tebben, a faculty member for 18 years in the department of chemistry at Presentation College and currently teaching at Northern State University, has introduced four new testing and assessment methods in her various courses:

1. Written reports on laboratory experiments. The instructor divides experiments into categories, according to the skills she wants to evaluate, such as presentation of data, mathematical calculations, and drawing conclusions from data. At the beginning of the semester, she works with students in class on each of these skill categories and gives them short assignments in the context of specific experiments. Then students are given a selection of experiments from which to choose, one in each skill category, and write a long report. Those experiments not covered in the long report carry a shorter assignment.

Students enjoy having a choice of experiments to write about, and the instructor enjoys not having to grade 60 versions of the same experiment report. She has used the system, developed from the writing-across-the-curriculum movement, for 10 years in freshman- and sophomore-level chemistry courses for both general education students and majors. Science colleagues with whom she has discussed the system seem interested, although Tebben knows of none who have adopted it.

2. Review of laboratory experiments. In an attempt to get freshman and sophomore science students to better observe and record experimental results, Tebben divides laboratory sections into small groups of four to six. Each group is responsible for reporting to the class as a whole a review of two or three experiments during the semester. Using an evaluation rubric of the instructor's design, the rest of the class evaluates each review, which is then compared to the instructor's assessment and to a self-assessment undertaken by the performing group as well. The grade for the lab review is the average of all three assessments. One of Tebben's colleagues has adapted this technique for her science classes.

3. Action tests. Students are required to find a way to demonstrate a concept or principle to illustrate some law of chemistry or physics, such as Newton's Laws of Motion, movement of air in tornadoes, or atomic structure. The instructor provides the criteria, such as specifying that the solar distances must be illustrated by concrete materials in which distance and size are proportional to the actual sizes and distances of the bodies assigned from the solar system. Students (individually or in groups) must supply a diagram and written explanation of the model to the instructor and may also be required to demonstrate the model to the class. Evaluations of these models are given by the professor and, if a presentation is made, by the class. If students worked in groups, an assessment of the group process is included.

Although she has used this method only in her physical science for teacher education course, she would use it in general education and major science courses as well if she taught them. Her colleagues in teacher education often use similar activities in their classes.

4. Writing questions for the test. In her physical science for education majors course, Tebben provides her students with her objectives for a unit of study. After appropriate activities and class discussion, students decide on the main concepts for the unit. In groups of four, they write original questions meant to evaluate their understanding of these concepts; they are directed to focus particularly on questions that will test higher-order thinking skills and then have to answer them.

In subsequent classes, the instructor shuffles these questions among the groups so that one group is answering and evaluating their answers to another group's questions. Their grade is a composite score of the quality of their questions, the quality of their answers, how well one group's questions were understood by another group, and the groups' critiques. Tebben finds that she can really gauge students' depth of understanding through this method. She has employed it in algebra, chemistry, and physical science for teachers courses.

At most institutions, says Tebben, "we do not often discuss our teaching and assessment methods with our colleagues, and so miss wonderful opportunities to share ideas."

"ADVANCE" COPY OF EXAM WITH KEY INFORMATION MISSING

Joel Tellinghuisen, Department of Chemistry, Box 1668 B, Vanderbilt University, Nashville, TN 37235; TEL: (615) 322-4873; FAX: (615) 322-4936; E-MAIL: tellinjb@ctrvax.vanderbilt.edu.

Courses Taught

- Introductory Chemistry
- General Chemistry
- Statistical Thermodynamics
- Quantum Chemistry II
- Data Analysis

Description of Examination Innovation

Joel Tellinghuisen generally includes a mix of medium-length problems, short-answer questions, and multiple-choice problems on his chemistry exams. In the fall of 1992, he used an exam that he distributed to the class the night before—except that this preliminary version was missing key information such as parts of sentences, quantities, and the names and formulae of substances (or parts thereof). He encouraged the students to work together to prepare for this exam in any way they pleased.

His goal was to get students to focus their effort on understanding these problems by recognizing how the missing information was needed to complete the test items. In a sense, he gave them a blueprint for the test.

The results of this experiment were mixed: Many students liked the format, but others "hated" it, says Tellinghuisen. Because he provided a preliminary version of the exam, he made the test more demanding than usual, both in sophistication and length. Some students simply were not well enough prepared 15 hours before the exam to make much use of the advance copy. In any event, the class average was in the mid-50s, when Tellinghuisen had expected an average around 80 percent.

Tellinghuisen plans to try this experiment again. He has considered giving students more than one night to study the advance version, but he also worries that some students will seek help from others besides their peers, which he has prohibited, such as TAs or tutors.

Tellinghuisen's colleagues, who have examined the preliminary test, have commented on how remarkable it is that an exam of this nature, one that "seems" to "give away the score," can nonetheless be substantial in its demands on students. The instructor is unaware of others adopting his method.

COMBINATION GROUP/INDIVIDUAL EXAM

Rob Thompson, Department of Chemistry, Oberlin College, Oberlin, OH 44074; TEL: (216) 775-8305; FAX: (216) 775-6682;
E-MAIL: robert.q.thompson@oberlin.edu.

Courses Taught

- Analytical Chemistry
- Trace Analysis
- Forensic Chemistry, "Chemistry and Crime" (for nonscientists)
- General Chemistry lab
- Chemical Information

Description of Examination Innovation

Having introduced student study groups some time ago in his analytical chemistry class, Rob Thompson has decided to employ study groups during the past year as part of his examination process in analytical chemistry and in a chemistry course for nonscientists. First, students take and complete an in-class or out-of-class examination on an individual basis. After the exams have been collected (but not graded or returned), the instructor gives the same exam to each study group to be completed as a group. The groups (usually four or five students) may use their notes, textbook, and as much time as they wish. The overall grade is based on 75 percent of the individual score plus 25 percent from the group score, resulting in overall gains in exam grades of 5–10 percent, due to group scores of between 90 and 100 percent.

With this gain, students are bound to like the new format. But the instructor values the practice because it forces his students to look back once again at the exam and to relearn the material. Peer discussions appear to be effective and nonconfrontational, permitting students to see what mistakes they made and how they made them. In addition, the instructor gets less blame for "bad questions" or "bias." Both the higher-scoring and lower-scoring students benefit, with the better students acting as teachers. Most important, at the end, everyone feels successful, and cooperation is shown to pay off.

In written comments, students give the following reasons for liking the group quizzes: (1) They could discuss the material with their peers; (2) they learned more; and (3) their grades improved. Reasons for not liking group quizzes include the following: (1) It was painful discovering wrong answers; (2) some groups did not work together well; and (3) it was hard to find the time to meet. Faculty colleagues have not made many comments about this innovation, unique at least at Oberlin College.

In arranging groups at the beginning of the semester, Thompson takes pains to include a range of abilities in each.

TAKE-HOME EXAM

Rosemarie DePoy Walker, Department of Chemistry, Metropolitan State College of Denver, Campus Box 52, P.O. Box 173362, Denver, CO 80217-2836; TEL: (303) 556-2836; FAX: (303) 556-4941; E-MAIL: walkerr@mscd.edu.

Courses Taught

- General Chemistry I and II
- Intermediate and Advanced Inorganic Chemistry
- Survey of Physical Chemistry

Description of Examination Innovation

For the past 6 years, Rosemarie Walker has been giving her chemistry students take-home examinations. Approximately one-third of the topics in General Chemistry II are covered by such exams and one-half of the topics in her survey of physical chemistry. She uses the take-home format for the following reasons: (1) to preclude having to provide large amounts of data for examinations (i.e., thermodynamics tables, tables of acid–base dissociation constants); (2) to be able to ask questions involving graphing; and (3) to improve the readability of student responses to essay questions.

Walker's out-of-class assessments are slightly more difficult and longer than in-class exams, observing her rule of thumb: If it takes her more than one-third of the time the student has to take the assessment to make the key, the assessment is too long. Questions focus on higher-order skills and interpretation of data rather than just content.

Students' responses, according to Walker, are "an odd mixture of regret that they spend so much time on the assessments, and pride in the fact that they feel their knowledge of that topic is far superior to their knowledge in other areas," given how much time they spend on the take-home exams.

Response from faculty colleagues has been mixed, a main concern being with the potential for student cheating. However, Walker uses a signed "Statement of Integrity" from students (affirming that the work in the exam is their own) to minimize cheating and also finds she can usually identify cheating via "uncommon errors." In the end, she doesn't worry much about this issue, because the comprehensive in-class final will be difficult for the student who has not done the work along the way.

Another change of Walker's has been to reduce time pressure by allowing students to work on an in-class examination during two nonconsecutive periods in her course. The way this works is this: Walker passes out an exam in one class hour, allowing the students to work on it. At the end of the class hour, she collects their exams and then returns them for completion during the next class hour.

Students are obligated on their honor not to take any notes from the exam, but they may, of course, commit questions to memory and study these again before the next exam period. And students may change answers during the second session. Although she keeps no statistics on how many students change answers, her feeling, from glancing at assessments between periods and listening to students, is that some modify answers, and many answer questions they left blank, but only students who have not understood the material well change answers entirely.

Walker has used this strategy for the past several years in her advanced inorganic and survey of physical chemistry courses. In addition, she provides a "Useful Information Sheet" during all her exams, containing formulae and commonly used constants because, as she tells her students, she doesn't attach much value to memorizing facts. "It is far more important for a student to know how and when to apply an equation than to know that equation by heart," she says. Walker also has her physical chemistry students conduct anonymous peer reviews of one another's formal lab reports. In the future, she plans to add a group presentation component to laboratory grades.

TRYING A LITTLE BIT OF EVERYTHING

Theresa J. Zielinski, Department of Chemistry, Niagara University, Niagara, NY 14109; TEL: (716) 286-8257; FAX: (716) 286-8254;
E-MAIL: tjz@niagara.edu.

Courses Taught

- Physical Chemistry

- Introductory Chemistry (for nonscience majors)

Description of Examination Innovation

When Theresa Zielinski first came to Niagara University over 15 years ago, she taught in a typical didactic method. As she recalls, her notes became her students' notes in the same way her teachers' notes had become *her* notes back in the 1960s. She measured student achievement in those days by standard exams, consisting of problems similar to those assigned for homework or presented by the instructor in class. "There were so many physical chemistry books around," she reports, "that it was not difficult to find a wide selection of problems."

Given her teaching load, at first Zielinski didn't have time to invent new problems. Yet, student performance on these standard tests was mediocre. This implied, she reasoned, either that "I was not a good teacher, the students were not working hard enough, the exams were too hard, or the students were not talented enough, especially because of their poor math skills."

At this point, in 1981, she began to experiment with various exam styles and options. Each mode was given several semesters of trial, and sometimes two or more exam types were used simultaneously. She coupled take-home with open-book exams, unlimited time with open book, and limited time with index card references. She even repeated certain exams, once with index cards, once open book.

Along the way she tried giving students a second chance at an exam of the same level of difficulty, only to find that they did not do much better on the retest. On the contrary, students indicated that because they knew they would have a second chance, they were not studying as hard for the first exam as they might. Even with open-book or index card references, student scores did not improve.

She persisted in her desire to find a way to change her students' study habits and provide them with a better learning environment. Having students go to the blackboard increased their participation in class. Computers in the lab also helped, but still exam scores in the lecture class were low. In time, she handed out hard copy of her transparencies to reduce students' note taking during class and to get them to think more. But even this did not improve exam scores.

In 1990, Zielinski started attending teaching workshops and seminars, where she learned about the Perry model of intellectual development of college students. In 1992, she attended an NSF-sponsored workshop on critical thinking in the sciences conducted by Craig Nelson in Dayton, Ohio. After these exposures, she switched over to a sequence in which her students got detailed study guides for each chapter of the text, prepared their own notes for each chapter by responding to the questions in the guide, and then in groups of four or five discussed the chapter in class. In this setting, blackboard work became routine.

At first, Zielinski collected and read everything that the students wrote for each chapter. Later she only randomly checked their notes. By monitoring interstudent question-and-answer exchanges, blackboard work, and student–teacher question exchanges, she began to do some nontraditional assessment. She learned to assess students' in-class work when she had them write in purple or green ink during class on notes they had taken before class in black or blue ink.

This finally resulted in better performance on exams, especially in students' approaches to problem solving. She now provides exam support, permitting students to bring their text and any study materials to an exam. And she has eliminated most curving for final grades. Less material is covered and, as a result, American Chemical Society scores have not improved, but in-class activity and scores on instructor-generated exams are substantially better.

Zielinksi has the advantage of extremely small classes (5 to 10 students). The results of all these improvements, in her words, are the following:

> First, students are making real effort to integrate the material for themselves. Second, they are no longer intimidated by large problems containing lots of data. Third, their exam scores are greatly improved, averaging above 80 percent. The quality of their answers has also improved, and they seem to appreciate the challenge of a difficult question and to enjoy the sense of accomplishment in a well-crafted answer.

ORAL EXAMINATION OF INSTRUCTOR BY STUDENTS, INDIVIDUALIZED ECLECTIC EXAMINATION

Uri Zoller, Department of Science Education–Chemistry, University of Haifa–Oranim, Kiryat-Tivon 36006, Israel; TEL: (972)-4-8242740/9838836; FAX: (972)-4-8345069/9832167.

Courses Taught

- General and Inorganic Chemistry
- Introduction to Modern Organic Chemistry

Description of Examination Innovation

As part of his effort to foster question asking and higher-order cognitive skills in freshman chemistry courses, Uri Zoller uses a unique testing strategy called "The Examination Where the Student Asks the Questions" (ESAQ), in place of a traditional midterm examination or (in written form) as part of a term examination.[16] The ESAQ is an oral examination in which the instructor is

---

[16] This description borrows from Uri Zoller, "The Examination Where the Student Asks the Questions," *School Science and Mathematics* 94, no. 7(November 1994): 347–349. The author has published extensively on testing and will supply a bibliography on request.

examined by students using written questions they have prepared at home. Each student prepares two to three course-relevant questions, one of which is used by the student in the oral examination session. After the examination session, all student questions are submitted to the instructor for inspection and grading. Two to five student questions not employed during the oral exam session are selected by the instructor and redistributed to students as a take-home examination to be completed individually and submitted after a day or two to the instructor for a grade.

Other variations on the ESAQ are possible, says Zoller, such as having students evaluate and grade (with guidance from the instructor) one another's take-home exams, as well as self-grading.

In the instructor oral examination session, Zoller handles approximately 20 questions in one and a half class periods with lively class discussion. Because he emphasizes higher-order cognitive skills (HOCS) and open-ended questions in his teaching, Zoller finds students' questions to be responsive in this respect and, quite often, on the HOCS level. Two examples of his students' questions follow:

> Compound *I*—C_6H_{12} underwent ozonolysis followed by treatment with Zn/HCl in aqueous solution. The two compounds obtained as products were CH_3CHO and $CH_3CH_2COCH_3$. What is the structure of *I*? Rationalize and explain your answer.

> The dialdehyde OHC-CH=CH-CHO was treated with molecular Br_2 to obtain the dibromide, the addition product of the bromine to the double bond. Will the product(s) obtained be optically active? Rationalize and explain your answer.

Students respond positively to the ESAQ, often for different reasons: Some enjoy challenging the instructor, whereas others are grateful not to have to take the traditional "difficult" midterm examination. A substantial number of students enjoyed the change from the traditional algorithmic, lower-order cognitive skills (LOCS) to the HOCS-oriented ESAQ in their exam experience.

In his organic chemistry course, mean scores of students on the midterm ESAQ questions, answers to selected questions on the take-home exam, and the traditional oral final examination were 77.2, 79.7, and 71.1, respectively. These and other research-based results make Zoller confident that this exam format, including the student preparation and related work after the "instructor's oral," results in substantial gains in student understanding and higher-order thinking capability.

In another innovation, Zoller assigns, in place of a final exam, an Individualized Eclectic Examination (IEE), a HOCS evaluation device based on each student's individual midterm or final project.[17] The multiformat examination

[17] Described in U. Zoller, "The IEE-an STS Approach (A Course Format to Foster Problem Solving)," *Journal of College Science Teaching* 19(March/April 1990): 289–291.

consists of written questions, problems, tasks, and ideas, suggestions, and opinions to be developed and rationalized, as well as experiments and simulations to be designed. The IEE by definition has no one, predetermined, standard format. It consists of desirable elements selected from different testing methods. The exam can be administered in class or given as a take-home.

In a typical, senior undergraduate general science class, Zoller's course requirements include, among other components, a research-oriented, written and orally presented final project on a topic selected by the student and an IEE. The projects involve library, laboratory, and experimental fieldwork, and the individualized IEE is administered to all students in open-book/notes format during the 2-hour final exam session. Most of the IEEs consist of three to four questions such as the following projects and corresponding sample IEE questions:

- Project: The effect of tap water, saltwater, and Fraser River water on plants. Question: Based on your knowledge, understanding, and judgment, and taking into consideration the obvious existing constraints, what are the three most important parameters (pollutants) that should be constantly monitored in the Fraser River? Explain and justify.
- Project: Caffeine consumption from soft drinks; the placebo effect of caffeine in cola drinking. Question: In your project you stated, "It is safe to say that caffeine's effects may be more serious than the general public is willing to admit." Based on the analysis of your own data and findings, what would be your recommendation(s) concerning caffeine consumption? Provide "hard evidence" to justify your recommendations.

The IEE helps the instructor to evaluate both problem-solving and decision-making skills, skills in analytical, synthetical, and, most importantly, evaluative and higher-order thinking. It encourages the problem solver rather than the exercise solver in students, and it incorporates each student's particular interests. Students respond enthusiastically to the IEE, saying that this type of examination is relevant to their needs and interests. The IEE also reduces students' test anxiety. Most of the faculty who learn of the IEE respond favorably to the idea, but it is not surprising that some express reservations with respect to the effort and time required by the instructor. Although it takes Zoller approximately one-half hour to prepare each IEE, he says the benefits of using the IEE outweigh by far the added preparation time.

CHAPTER 5

GEOLOGY

CHALLENGING MULTIPLE-CHOICE QUESTIONS, COMBINED INDIVIDUAL/GROUP EXAMS

Ann Bykerk-Kauffman, Department of Geosciences, California State University–Chico, Chico, CA 95929-0205; TEL: (916) 898-6269; FAX: (916) 898-4363; E-MAIL: akauffman@oavax.csuchico.edu.

Courses Taught

- Introductory Geology
- Structural Geology
- Earth Science
- Historical Geology
- Field Geology

Description of Examination Innovation

When Ann Bykerk-Kauffman began teaching a large section (140 students) of an introductory geology class for nonscience majors at California State University, Chico, in 1991, she was determined both to make her course interesting to recruit students to the major and, particularly, to figure out a way to have her students learn from their exams.

Her father, Roelof J. Bijkerk, a psychology professor at Grand Valley State University in Michigan, shared with her his considerable expertise in the design of machine-graded, multiple-choice exams that test deep-level learning and high-order skills, and not just vocabulary. He pointed her in the direction of the classic literature on test-item construction and advised her to give her students the opportunity to retake their exams; but she wanted to experiment with cooperative exam-taking as well.

After several years of tinkering, Bykerk-Kauffman came up with a technique that she and her students believe is fair, does not create extra work for the instructor, does not take more class time than a standard exam, and, most important, makes the exam a less stressful, more thorough learning experience. Her system is based on the "Team Effectiveness Design" developed by Jane Mouton and Robert Blake. This is how it works:

Bykerk-Kauffman gives two lecture exams during the semester. (An additional exam on rock and mineral identification is given in lab.) Instead of a final exam, students do a final project. Both lecture exams consist of 50 challenging multiple-choice questions with five options each. A correct answer to an exam question will typically require a deep level of understanding, she explains, the ability to apply knowledge to specific situations, and/or the ability to synthesize disparate pieces of information into something new.

The exam and preparation for the exam take up three 50-minute class sessions. During the first session, students work on a practice exam in the small cooperative learning groups to which they have been assigned since the beginning of the semester. Practice exams are made up of actual questions from old exams. As the students work the practice exam, she writes the "correct" answers to the questions on the board, 10 every 10 minutes, and she circulates through the room answering student questions. (Undergraduate teaching assistants are on hand as well.) In her experience, active involvement in a practice exam is a much more effective way for a student to prepare for a test than passively listening to a professor reiterating previously made points.)

During the second session, students take the test individually in the standard manner, recording their answers on computer sheets. The usual precautions against cheating are taken.

In the third session, students retake the test in their learning groups. Consultation between groups is allowed, but not the use of books or notes. Again, students record their answers on individual computer sheets; and they do not have to defer to the group answer if they think they know better. Because this "going-over-the-exam session" involves points, students are more motivated than they would otherwise be to learn from their mistakes.

The exam grade for each student consists of the sum of the grades from the individual and group exams. There is no penalty for scoring lower on the group exam than on the individual exam. If it happens that some smooth talkers are able to talk their teammates out of their correct answers, Bykerk-Kauffman simply doubles the score on the individual exam. (The same is done for students who miss the group exam.) Because her exams are "difficult," she generally adds 5–10 points to each student's earned score, bringing the average score up to around 75.

Because test-item construction is so time-consuming, Bykerk-Kauffman never lets her students take the exams home with them. Since she rarely uses exam questions provided by the textbooks (which she finds "abysmal," as they generally test knowledge of vocabulary or trivial facts), she is ever in search of good questions. Therefore, she allows students to earn extra credit points by submitting questions, allowing 5 points (out of 1000 possible points in the course) for each *usable* question.

Ever since she began using group exams, she has observed with delight the "vigorous debates" that occur during group discussions. But she warns that instructors cannot spring group work on students *only* during exams. Her group exams are embedded in a class that involves a great deal of cooperative, small-group activities. Thus, the group format is not unfamiliar to them when a test comes along. Students are taking their tests together with other students with whom they have already worked a great deal, especially in lab. She doesn't believe group exams would work as well were this not the case.

The system is popular with students, who report that they learn a great deal from the process and actually enjoy the group exam, even though the questions are difficult. The only complaint she has received from large numbers of students is that the practice exam seems to be much easier than the real exam. She believes this is a misperception. Students perceive the "real" exam to be more difficult because of increased stress and lack of help from fellow students when they take a comparable exam individually.

Several of Bykerk-Kauffman's colleagues at California State University, Chico, have expressed interest in technique, but only one so far (no longer teaching because he has become a dean) has adopted it. Bykerk-Kauffman has also heard from a colleague who has had great success with a modified version used in nutrition classes.

Ann Bykerk-Kauffman has a collection of written student responses to her technique and examples of her exams which she is willing to share with other faculty who write to her.

GROUP QUIZZES

P. Thompson Davis, Department of Natural Sciences, Bentley College, Waltham, MA 02154-4705; TEL: (617) 891-3479; FAX: (617) 891-2328; E-MAIL: pdavis@bentley.edu.

Courses Taught

- Introductory Geology I and I
- Environmental Geology

Description of Examination Innovation

Three years ago, P. Thompson Davis began allowing students to take their mineral and rock quizzes in groups. The quizzes simply consist of identifying the specimens by name, but to do so correctly requires a number of hypothesis-testing procedures. Students work in groups when studying the specimens, so he decided to let them remain in their groups for the quizzes. He considers group testing a "final step to cooperative learning" in the introductory geology laboratory, which is usually small (less than 30 students). About 30 percent of the final grade is derived from group-based lab work.

Davis has observed many more positive features of student behavior since he began group testing: working together, arguing and debating points, reaching consensus. Results, he believes, are more positive than negative. There are some "freeloaders," but not many, and students say they like the approach.

QUESTIONS ABOUT SCIENTIFIC WRITINGS

Pascal de Caprariis, Department of Geology, Indiana University–Purdue University, 723 W. Michigan Street, Indianapolis, IN 46202; TEL: (317) 274-7484; FAX: (317) 274-7966; E-MAIL: pdecaprr@iupui.edu.

Courses Taught

- Environmental Geology (introductory service course)

Description of Examination Innovation

For the past 2 years, Pascal de Caprariis has been using an alternative assessment in addition to the standard multiple-choice examination. He assigns five 2-page, double-spaced papers on short articles that are relevant to the course material. Students must respond to a statement and question or questions about the article. For example, upon reading the introduction to Great Possessions: An Amish Farmer's Journal, students are asked to express an opinion about the paradox of the seemingly old-fashioned farming techniques of an Old Order Amish community against today's standards, defended by author and farmer David Kline. The students' work is graded on grammar and on the organization of their ideas. (The answers must be coherent as well as grammatically correct.)

The instructor ties his two assessment techniques together by including questions from their readings on the in-class, multiple-choice tests. But he doesn't tell students about these questions, because he doesn't want them to memorize the articles. He presumes that if they have done the papers, they can answer the questions.

The first time de Caprariis tried this approach, some students reacted negatively to being graded on grammar in a science class. His response was that "Life is an English class," and they ought to get used to being evaluated on how they expressed themselves.

The instructor says that without the assistance of an English major, also a freelance writer, who is paid by the department to assist with grading, he would never be able to use these papers with his class of 125. The chair's willingness to fund the extra grading assistance demonstrates his support for de Caprariis's technique. Most recently, another faculty member used de Caprariis's approach with a slight modification: Students could choose their own topic for the last paper of the semester. This was popular, and de Caprariis will probably repeat this feature.

TUTORIAL ORAL EXAMS

Richard H. Fluegeman, Jr., Department of Geology, Ball State University, Muncie, IN 47306-0475; TEL: (317) 285-8367; FAX: (317) 285-8980; E-MAIL: rhfluegeman@bsuvc.bsu.edu.

Courses Taught

- Stratigraphy

Description of Examination Innovation

Fluegeman teaches stratigraphy in a nontraditional tutorial format. Students meet in pairs with the instructor for a 1-hour session each week of the semester (14 total). The topic of the tutorial is a chapter in the textbook. Students are asked to bring questions to the tutorial from that chapter. Although the students are encouraged to study in pairs, they are under no obligation to do so, and, in fact, few do. The instructor assigns grades individually by evaluating each student's preparation. The instructor also uses the time to test students' knowledge of the assignment, which he evaluates at the time.

Fluegeman requires no written examinations in the course. Overall ability is measured by the tutorials, two theme papers, and one poster presentation. Each theme paper is assigned to a topic appropriate for that portion of the course under discussion. Different from a term paper, the theme topic is selected by the instructor and deals with a stratigraphic issue or a stratigraphic theory and requires the student to take a stand and defend that position using stratigraphic data. Each tutorial receives a grade on a 10-point scale, and in total they are worth one-third of the semester grade. The themes and poster are worth one-third, and the lab is worth one-third as well. Such a format permits the instructor not only to evaluate students' factual knowledge in stratigraphy but also their ability to apply knowledge to the topic—particularly important in a course with a laboratory component.

Each student's poster session involves the description of a depositional environment using as little text as possible, relying on illustrations from the literature to get the concepts across. The presentation is limited to a 24" × 26" posterboard. The posters are exhibited in the department for 2 weeks before being graded. Comments from faculty and graduate students are welcome.

The oral, theme, and poster, in combination, have, in the instructor's opinion, improved all students' understanding of the subject. In addition, students find the new assessment practices to be less intimidating than the tradition midterm and final. Students accept their grades as a result of their efforts and of their mastery of the material, and not as the fortuitous (or not so fortuitous) consequence of the set of questions selected for any one examination by their instructor. Fluegeman points out that so far, he has not had a student challenge a tutorial grade. The instructor says students have appreciated his assessment methods, particularly their individual contact and being able to comment on the scientific work of well-known geologists. Indeed, he has been told many times that this class is the first in which the student found reasons to read the entire textbook.

Fluegeman has been using this combined assessment strategy for 3 years. Due to increasing enrollment, he feels it is unlikely that this method will be used in other classes. The largest group he has ever used the method with is a section of 14 students, and he thinks sections with more than 15 students could not effectively utilize this method.

TAKE-HOME EXAMINATIONS

Paul K. Grogger, Department of Geology, University of Colorado, Colorado Springs, CO 80933-7150; TEL: (719) 593-3136; FAX: (719) 593-3146; E-MAIL: pgrogger@brain.uccs.edu.

Courses Taught

- Physical Geology
- Historical Geology
- Geomorphology
- Environmental Geology

Description of Examination Innovation

Paul Grogger employs take-home field exams in all his courses. Some require in-field decisions; others need sample collecting and subsequent laboratory analysis. The fieldwork takes place at various locations throughout the Pikes Peak region and at numerous stream and lake locations. All exams are design to test the ability of the students to see, determine, and communicate their knowledge about the processes and concepts of the course they are completing.

Grogger started using this method in his more advanced classes. After determining that the method might work in introductory classes, he adapted it to the introductory classes he teaches. Because in-field testing has been introduced, he has been able to increase the students' time in the field to more than one-half of class time. According to course evaluations, students enjoy this fieldwork and believe it contributes to further learning. The only problem he has had is that some disabled students have difficulty completing fieldwork. Two of eight instructors in the geography and geology departments have started using the method, though less extensively than Grogger.

MINERAL IDENTIFICATION LAB PRACTICAL

Kurt Hollocher, Department of Geology, Union College, Schenectady, NY 12308; TEL: (518) 388-6518; FAX: (518) 388-6789;
E-MAIL: hollochk@gar.union.edu.

Courses Taught

- Mineralogy
- Physical Geology
- Igneous and Metamorphic Petrology
- Geochemistry

Description of Examination Innovation

For 6 years, Kurt Hollocher has been handing out the bulk of his mineralogy lab practical on the first day of class. The exam consists of a box containing 65 minerals that each student must identify by the end of the term. Students are allowed to use any of the analytical techniques available in the department except quantitative chemical analysis. They can use the properties of color, hardness, crystal form, and fizzing in acid, among others. In addition, students are taught and may use optical properties of the crushed mineral grains under the microscope; they can use density measurements and the X-ray diffraction machine (without the automatic search function). The latter is used on only 10–20 percent of the minerals, usually as a final check, since the students quickly learn that X-ray diffraction can be a difficult technique to apply unless one has already limited the choice of possible minerals.

Students have always been enthusiastic about the exam, says Hollocher, spending long hours in the lab, of their own free will, identifying the minerals. The exam personalizes mineral identification, since it rapidly becomes clear to the students that the quality of the end result is entirely up to them. If they put in the hard work and thought necessary to identify the minerals in the set, then they will do well. They have a whole term to do the work and are repeatedly reminded of their task every time the class reviews a different mineral group or works on a new identification or measurement technique. Through peer pressure, the students seem to set a steady, reasonable pace for work.

Students hand in not only the names for each of the mineral samples in the test, but also the results of any tests made. The grade is based partly on the thoroughness and correctness of the properties used to identify the samples. Partial credit is given for incorrect but reasonable answers, considering the data that were collected. Tenth-of-a-point deductions are made when the answer is correct or reasonable but additional easy measurements would have allowed a more definitive answer.

Although every student completes the test, not all student do the *same* amount of work. For example, one student might identify quartz simply by comparison to other samples seen in class, which is a valid but incomplete argument for correct identification. Another student might also measure hardness, physical form, some optical properties, and density. On Hollocher's exam, the former student would receive a few tenths of a point off for not having made a sufficiently strong case for identification.

"The students learn more about mineral identification than they ever did in my earlier mineralogy courses," Hollocher adds, and the exams are "surprisingly easy" to grade.

The students get to keep their minerals, too, which they like.

THE "TWO-INK" ASSESSMENT

Gene W. Lene, Department of Earth Sciences, St. Mary's University, One Camino Santa Maria, San Antonio, TX 78228; TEL: (210) 436-3235, Ext. 1434; FAX: (210) 436-3500.

Courses Taught

- Environmental Geology
- Geology of Energy Resources
- Essential Elements of Life–Earth Sciences
- Mineralogy
- Geomorphology
- Structural Geology

Description of Examination Innovation

In most of Gene Lene's introductory courses, tests include some combination of multiple-choice, matching, fill-in-the-blank on diagrams, and short-answer essay questions. To reduce test anxiety, to encourage organized note taking, and to deemphasize rote memorization, Lene adopted the following simple test procedure in the spring of 1994.

1. Each student is provided with a green-ink pen with which to take the test in the usual manner.
2. When satisfied that they have done their best, students trade their green-ink pens for red-ink ones. With the red-ink pens, they are allowed to change any answer while consulting their personal notes.
3. Correct "green answers" are given full credit. Correct "red answers" are given one-half credit. "Red answers" are considered final. "Green answers" that have been superseded by "red" ones are ignored, even if they were correct. In this way, students learn to recognize what they know as well as what they don't know, and the test becomes a far more valuable learning experience than those following a traditional one-time format.

It is too early to determine the long-term results of this innovation. So far, Lene reports that initial fears of grade inflation have proved to be unfounded; there seems to be only a modest improvement in grades. Students, however, have reacted "very favorably," reporting that they are much more comfortable with the tests. Upon consulting their notes, many have found that they actually need to change few answers. Colleagues with whom Lene has shared this technique have expressed interest in it. However, most of them use computer-graded answer sheets and could not use the technique.

The procedure is simple, the pens are inexpensive, grading time and effort remain the same, and little change in test procedure or test structure is required. If student morale and self-confidence are only slightly improved, says Lene, the procedure is still well worth doing.

BOTTOM-LINE DISCLOSURE AND ASSESSMENT

Edward B. Nuhfer, Office of Teaching Effectiveness and Department of Geology, University of Colorado at Denver, Campus Box 137, P.O. Box 1773364, Denver, CO 80217-2678; TEL: (303) 556-4915; FAX: (303) 556-2678; E-MAIL: enuhfer@carbon.cudenver.edu.

Courses Taught

- Environmental Geology (introductory)
- Engineering Geology (graduate)

Description of Examination Innovation

In his "Bottom-Line Disclosure and Assessment" system, Edward Nuhfer makes certain his students know what is expected of them in his class. From his entire bank of test questions from years of teaching (ordered in the sequence topics are covered), the instructor creates what he calls a "monster exam" that covers the entire course in the form of test questions. Each student gets a copy of the "monster exam" for his/her own files and responds to the exam as a "knowledge survey."

The survey is given on or near the first day of class. Students indicate on a scannable answer sheet whether they could answer any of these questions now, without taking the course ("A"); whether they could partially answer the question at this point, or would know where to go for the information within a 30-minute time frame ("B"); or if they do not know either the answer or where to go for it ("C"). Students hand in their answer sheet and get to keep the quiz questions. Students are cautioned that they can be asked to answer, for a quiz grade, any three questions for which they marked "A." This statement is given to ensure that students give serious thought to each question and to their ability to answer it. Nuhfer has never yet "invoked the threat," but students say it keeps them honest.

The exercise, says Nuhfer, shows what kinds of preparation students bring to class. If the class has common deficiencies, this becomes the time to discover them, rather than a month later, during the first exam.

When the exercise is exactly repeated at the end of the course, instructor and students can verify the amount of "value-added" knowledge that has been provided by the course. Nuhfer writes in an article about his system that "indirect methods of 'evaluation' (colleagues' opinions, surveys of students' satisfaction) are important, but these are not actual measures of value-added knowledge."[1]

Students like the procedure. "Disclosure," Nuhfer reminds us, "is an important factor that contributes to student satisfaction." It is also a way to provide repetition in confronting material. Disclosure and monitoring of progress are good ways to "manage" a course, says Nuhfer. In fact, the National Veterans Training Institute at Colorado University–Denver has adopted this knowledge-survey practice as a way to assess all of their professional skills development workshops. Since 1995, the core science courses at the institution have been assessed through use of a knowledge survey with great success.

---

[1] From a University of Colorado at Denver newsletter. A condensed version of Edward Nuhfer's "Bottom-Line Disclosure and Assessment" is printed in *The Teaching Professor* 7, no. 7(August–September 1993): 8.

A detailed description of "Bottom-Line Disclosure and Assessment" prepared for colleagues is available from the author to those who write him.

SHORT ESSAYS, TAKE-HOME FINAL

June Oberdorfer, Department of Geology, San Jose State University, San Jose, CA 95192; TEL: (408) 924-5026; FAX: (408) 924-5053;
E-MAIL: june@geosun1.sjsu.edu.

Courses Taught

- Introductory Hydrogeology
- Geology and the Environment

Description of Examination Innovation

For the past 5 years, June Oberdorfer has assigned a take-home final exam. The exam requires students to synthesize information from the entire course and includes material from a number of laboratory exercises as well. The exam presents actual field data from a small groundwater contamination site, and students are required to analyze the geologic, hydraulic, and chemical data to determine where and how fast the contaminants are moving and how best to clean up the site. The end report is similar to a consultant's hydrogeologic investigation.

In addition, starting 11 years ago, the instructor has administered short essay exams of 10 questions each that also require students to synthesize information from the course and, in some cases, to form their own opinions about the implications of geologic phenomena for human actions. About 60 percent of the questions are based on lecture materials, class activities, and films. The other 40 percent are taken from review questions at the end of the textbook chapters (an incentive for students to read the textbook).

Oberdorfer has introduced these innovations to counter students' tendency simply to regurgitate facts in a science course. She refuses to give multiple-choice tests (despite the heavy grading load of reading short essays), because she believes students "need to understand the relationships and processes that can only be shown in essays, and because they need the opportunity to express themselves in writing."

Many students are initially intimidated by the short-essay format, being more accustomed to multiple-choice questions. Usually, they perform worst on the first exam and then improve as they grow accustomed to the instructor's style of questioning. Grading, however, is laborious: Oberdorfer's students spend a total of 35–55 minutes on the 10 short essay questions, while the instructor takes 6–10 minutes to grade each question—60–100 minutes per exam.

UNTIMED OPEN-BOOK PARTIAL GROUP EXAM

Dexter Perkins, Department of Geology, University of North Dakota, Grand Forks, ND 58202; TEL: (701) 777-2991; FAX: (701) 777-4449;
E-MAIL: dexter_perkins@mail.und.nodak.edu.

Courses Taught

- Introduction to Geology
- Environmental Issues
- Mineralogy
- Petrology

Description of Examination Innovation

For 10 years, Dexter Perkins has been teaching a mineralogy course, taken by all geology and geological engineering majors, which requires mastery of a great deal of material that students have never seen before. "The volume and nature of the content of the course can be overwhelming," Perkins reports.

During that whole time, the instructor has been unhappy with the way exams worked. Traditional hour-long exams didn't allow enough time to delve into the interesting and significant aspects of mineralogy. If he asked his students "think questions" or "conceptual questions," the students (most of whom had never had to answer such questions before) simply couldn't handle them—especially under time pressure. If he asked the usual short-answer, multiple-choice questions, he felt he wasn't testing any more than the minutiae students had memorized. If he asked students to explain something, rather than just regurgitate factual information, he found their answers to be confusing, rambling, and poorly written.

For the past 3 years, Perkins has tried a new exam format: open-book exams with no time limit. In addition, he has made all or part of each exam a group effort. The goal of this approach is two-fold: to take the pressure off students, and to use exams as a learning device by asking difficult and significant conceptual questions.

Test items are open-ended, often with several different answers, depending on the approach taken. In some cases they are equivalent, says Perkins, to what an instructor would ask a Ph.D. student on a comprehensive exam, requiring lengthy, well-written answers. For example, one recent question asks for an explanation as to why cubic minerals have simple X-ray patterns, whereas triclinic minerals have complex ones. Another states:

> Copper can generate other X rays that aren't Ka radiation. Diffractometers, however, have filters to remove the other X rays. Why is it important that the filters

are used to give monochromatic radiation? Clearly explain your answer so that even a geophysicist or paleontologist can understand.

Students must demonstrate an understanding of how science works and how scientists think. Organization, logic, and good writing are also essential for full credit. Because of the length of the questions, most exams consist of no more than 6–10 items.

Class response so far has been quite positive. Although most students say they spend too much time on these exams (they are right, according to Perkins), they all say that they learn more doing the exams than from any other part of the course. Even if weaker students take a free ride, the group efforts work well, the instructor thinks, because he believes that the weaker students wouldn't have gotten anything out of traditional exams, either. From the harder-working students, he says, "The insight and depth of some of their answers is amazing."

Top students typically spend 3–8 hours on the exams. Weaker students (except those who just don't care) spend as much as twice that time. Perkins doesn't believe the tests require that much effort, but given the opportunity to use books and other resources, the better students aren't comfortable until they have checked out every possible source from which to add to their answers. In previous courses, they have been trained to concentrate on details rather than on the big picture. The instructor believes that only by doing more exams of this nature will they learn to trust their knowledge and an intuition based on that knowledge.

According to Perkins, concerns about how to grade group efforts keep his colleagues from adopting this method. But this did not seem to trouble the student who wrote on his evaluation of the group testing method: "Everyone pulled his/her weight well, and although the questions were often split between group members, all students were very conscientious about reporting findings and making sure everyone understood the answers found for the questions."

OPTIONAL ESSAY EXAM

Kenneth L. Verosub, Department of Geology, University of California, Davis, CA 95616; TEL: (916) 752-6911 or -0350; FAX: (916) 752-0951; E-MAIL: verosub@geology.ucdavis.edu.

Courses Taught

- Geologic Hazards
- "The Earth," Introduction to Geology

Description of Examination Innovation

For the past 20 years, Kenneth Verosub has taught Geologic Hazards, a two-credit, general interest lecture course. The class enrolls about 250 students, most of whom are not geology majors. Because his only assistant is a grader, exams in the course have consisted solely of multiple-choice or fill-in questions. He used the same format 3 years ago, when given responsibility for "The Earth," a three-credit, general education, introductory geology course for approximately 250 nonmajors.

In 1994, as part of a general self-examination of what he taught and how he taught it, Verosub decided to offer students in both his classes the option of taking an essay exam as an alternative. He had been toying with this idea for some time, because he was concerned that the usual multiple-choice/fill-in exams tested only one limited dimension of a student's mastery of a topic. Verosub had already tried to go beyond simple terminology and definitions in traditionally formatted exams over the years by including synthesis and analytical-thinking questions on these exams. (Some students complained, particularly when Verosub could not show them where the correct answers to these questions could be found in the book.)

As a prelude to offering the essay alternative, Verosub took 5 minutes at the end of one class and asked students to state briefly (and anonymously) in writing whether they preferred multiple-choice/fill-in exams or essay exams, and to give the reasons for their preference. The vast majority said they preferred multiple-choice/fill-in exams. The most commonly stated reasons for this choice were "Essay exams don't test what we know" and that "Grading on essay exams is too subjective."

The survey encouraged Verosub to offer the essay alternative, because it suggested the number of students who would take the essay option would not be unmanageably large. The survey also confirmed his belief that the most appropriate essay question would be a general one that could be answered in a variety of ways. Finally, it made him aware of the need to provide clear guidelines about how he intended to grade the essay.

Verosub's main goal in offering the essay option was to provide students with a means of demonstrating how they had mastered the material in the course. For him, this meant showing how well they understood its main ideas and broad concepts rather than demonstrating how well they could organize the terms and definitions into paragraphs. He concluded that in order to be consistent with this goal, the essay exam would have to be open-book and open-note.

Fourteen students took Verosub up on his essay exam offer (as an alternative to the midterm) in his geologic hazards course. For the final, 19 students took the essay exam. In his introductory geology course this spring, 34 took essay exams in lieu of the first midterm, 46 took it in lieu of the second midterm, and 49 took it in lieu of the final. Each time, about 60 percent of the "takers" were women. Twenty-three students took the essay on all three occasions, and 11 took it twice.

Each essay exam consisted of one general question. For example, the first half of the course on geologic hazards dealt with earthquakes. The midterm essay assignment on this topic was to write a reassuring letter to a young woman who had just moved from Boston to Los Angeles and was very nervous about earthquakes. To answer this question, students could draw on any one of several sources covered in the course, such as earthquake preparedness, building codes, causes of earthquakes, the pattern of faults in California, and lessons learned from recent earthquakes. For the final exam in the introductory geology course, Verosub drew upon the fact that four Southern universities had recently abolished their geology departments and asked the students to write a letter to a fictitious academic administrator in the South, making the case for students to have the opportunity to take an introductory geology course.

Although the situation-statements of the essay exam were quite brief, they were accompanied by additional paragraphs that elaborated on the question, offered advice on how to plan and write the essay, and explained how the essay would be graded. Verosub indicated in these instructions that he was particularly interested in essays that were creative, imaginative, and went beyond simple recitation of factual material readily available in the text or notes.

Grading the essays actually proved easier than the instructor had expected. Upon reading for range of response, he found that the essays always fell into three categories. In the middle were essays that answered the question with a series of examples drawn from the material in the course. At the top were essays with examples that were integrated into a more coherent or creative framework. At the bottom were essays that did not adequately address the question, or went off on some irrelevant tangent. For example, in response to the question about the importance of providing introductory geology courses to students, most students based their arguments on what they had learned in Verosub's course. A few students, however, were able to transform their own experiences (and their knowledge of the geology of the South) into an essay on what students at a Southern university might gain from such a course.

The gradations within categories allowed Verosub to produce a set of scores conforming more or less to the canonical bell-shaped distribution. At that point, he faced a difficult dilemma, namely, how should the mean of an essay exam be related to the mean of the corresponding multiple-choice/fill-in exam? This problem was especially acute, because those who took the essay exam were self-selected and almost certainly better as a group at writing essays than those who chose not to take the option. To say that the mean score on the essay had to be equivalent to the mean score on the multiple-choice/fill-in exam seemed grossly unfair to Verosub. Unfortunately, he didn't know how to objectively determine the differential between the two exams. In the geologic hazards course, where only about 5 percent of the students took the essay exam, he finally decided that a mean of about 45 out of 50 seemed right for the essay exam when the mean on the multiple-choice/fill-in exam was about 37. For the introductory geology course, where almost 20 percent took the essay, Verosub felt more comfortable, with a mean of 42 for the essay exam compared to 37 for the other.

Overall, Verosub views this experiment as a success and plans to continue it in subsequent years. Several students who took the essay exam sent him e-mail messages expressing their gratitude at being given this opportunity. Others used the course evaluation form to comment favorably on the experience. A few students who chose not to take the essay exam complained that essay-exam takers did not have to study as hard as they did. But these students did not explain their choice not to take the essay exam. Most of Verosub's colleagues have been generally indifferent to what he did, a few regarding it as too much effort to be usable.

Verosub says he's not sure the experience clarified his understanding of what he's doing when he gives an exam. In recent years, in fact, he has become increasingly confused on this point. Prior to giving the essay exam, he couldn't decide whether an exam tested his ability to teach, the ability of his students to learn, the ability of their parents to instill in them a commitment to learning, or the ability of their high school teachers to teach them how to learn. To this list, he can now add the ability of their English teachers to teach them how to write.

TAKE-HOME FINAL EXAM GIVEN FIRST DAY OF CLASS, ROLE-PLAYING SCENARIOS

Jamie Webb, Office of Faculty Development, California State University, Dominguez Hills, 1000 East Victoria Street, Carson, CA 90747; TEL: (310) 243-3387; FAX: (310) 516-4268; E-MAIL: jwebb@research.csudh.edu.

Courses Taught

- Historical Geology
- Earth Science

Description of Examination Innovation

In historical geology, a course covering the earth's history for the last 4.6 billion years, a comprehensive, in-class final examination didn't seem practical or sensible to Jamie Webb, geology professor at California State University, Dominguez Hills. So she developed a comprehensive take-home exam she uses to great advantage with her students.

The test is handed to students as early as the first or second meeting of the semester. Although only two pages long, the 300 points include lengthy questions. For example, students are to provide a detailed description of the organisms involved in major evolutionary changes and adaptations throughout the earth's history, and a thorough discussion of organisms affected by the extinctions at the end of the Paleozoic and Mesozoic Eras and the Pleistocene Epoch. Webb tells students that they may discuss and even work on the exams together, but they must submit answers written individually, in their own words.

"Handing out the exam on the first day gives students a good idea of the amount of work required for the class," says Webb. Frequently, 5–10 percent drop at that point.

The results of this exam have been "very good, certainly much better than an in-class test," reports Webb. Students frequently tell her they have a better appreciation for the connection of events over time when they complete the time chart in the exam, which includes mountain making, evolution, and shallow oceans. Answering such in-depth and comprehensive questions, for which some students tackle computer graphics or other illustrating tools, could not be included on a timed, in-class exam. Likewise, the question on extinction requires students to compare different extinction events in depth, reinforcing the idea that extinction, even mass extinction, is not a unique event, but has occurred frequently during the history of the earth.

Webb uses another innovative format in her earth science assignments: open-ended role-playing scenarios, in which students must "play out," in groups, a hypothetical situation, such as a futuristic prediction, 85 percent accurate, by a Cal Tech seismologist of a major earthquake in Los Angeles in 2 years. The seismologist has decided to let the government determine whether to inform the citizens of Los Angeles of this information. Participants in a meeting held by the mayor to make this decision include a real estate developer, an insurance agent, a home owner, an elementary school teacher, and a civil defense director. Each group must plan a presentation, lasting 30–45 minutes, in which each group member assumes a particular role and argues and supports his/her recommendation. Other scenarios include creating a real estate development plan and implementing earthquake preparation. Group members practice, helping each other develop arguments and make the presentation scientifically accurate and believable.

Webb grades the scenarios on content and balance of information. Key information, such as the potential drop in property values if the quake is forecast, is needed for a thoughtful and accurate presentation (which is more relevant to Webb than specific content). Generally, she gives one grade to each group, although she is considering changing this practice. She has also considered having the rest of the class participate in grading the presentations.

Students seem to like the scenarios, especially because of the reality of earthquakes in the region. Some even visited government officials to get different perspectives before giving their presentation. Webb recalls one particularly bright student who did minimal work in her class but excelled in the scenario because of the high interest and relevance of the topic.

In large classes, the students don't "act" the scenarios but research and write their answers in a take-home assignment that may appear in some form on a later test. "Generally, these written answers are two to three times as long as any other answer on an exam, and much better written and organized," says Webb. "Some student answers are worthy of consideration by disaster planners." She thinks these questions teach students about scientific process and give them an appreciation and general understanding of geology and how it fits into their lives.

A newly hired faculty member at Webb's institution has adopted many of her objectives, including the earthquake prediction and a similar landslide scenario, but, generally, her colleagues haven't held to a "less is more" philosophy—though colleagues in physics and other sciences have gone even further, adds Webb.

CHAPTER 6

PHYSICS

FREQUENT, UNANNOUNCED QUIZZES

Timothy Barker, Department of Physics, Wheaton College, Norton, MA 02766;
TEL: (508) 286-3975; FAX: (508) 285-8278;
E-MAIL: tim_barker@wheatonma.edu.

Courses Taught

- Extraterrestrial Life
- Cosmology
- Planetary Science (introductory course for nonmajors)
- The Nuclear Age (first-year seminar)
- Astrophysics
- Frontiers of Astronomy

Description of Examination Innovation

Many of Timothy Barker's classes now start with an unannounced 15-minute, five-question quiz on the assigned reading for the day, which is corrected immediately in class by the students themselves. They use red pens for grading and are encouraged by the instructor to take notes as well as make corrections. The three lowest quiz grades each semester (out of 12 total) are dropped to allow for absences.

Quizzes have transformed Barker's teaching. His students come to class prepared and on time. "I'm much less concerned about trying to cover the material now," says the instructor, "and I feel more relaxed about discussing tangential ideas." He says even on days when there isn't a quiz, when he starts off asking for student questions on the reading, several hands always fly up, a great difference from the "old days." In addition, more students come to his office before class with questions that help him decide what to discuss in class.

During the class quiz review, discussion about students' "wrong" answers gives them much more insight into the material than the lectures, says Barker.

His review of the student-corrected quizzes while recording grades gives him further feedback about student learning. Many students report that their reading skills have improved dramatically. The instructor also finds it easier to call on shy students, knowing they've come prepared and have answers to quiz questions written down in front of them. Analysis of class videos indicates that there is a much greater gender balance in class participation than before the quizzes.

The quizzes, worth 30 percent of the course grade, are used in all of Barker's classes. In Astrophysics and The Nuclear Age, with 8 and 15 students, respectively, he has students work in groups of two or three to devise a quiz (which the instructor reviews and duplicates) and then lead the class in correction and discussion of it.

For the last 5 years that Barker has used these quizzes, student reactions have been positive. The instructor also credits his students' improved exam performance to the frequent quizzes, as well as his own improved test-writing abilities.

After giving several talks at Wheaton about this technique, Barker knows at least two colleagues, one in French and one in physics, who now use it.

VARIANT OF SELF-PACED/MASTERY LEARNING

John Burton, Department of Physics, Box 71917, Carson Newman College, Jefferson City, TN 37760; TEL: (423) 471-3260; FAX: (423) 471-3502; E-MAIL: burton@cnacc.cn.edu.

Courses Taught

- General Physics

Description of Examination Innovation

For 15 years, John Burton has used a self-paced examination scheme in general physics. A series of quiz topics is outlined in the syllabus, involving all the concepts that are fundamental to the course. Each "quiz" (of which there are two levels) is actually only a genre of physics problems. C-level quizzes generally require the use of one basic principle of physics, whereas A-level quizzes involve more than one. To get a grade of C or better in the course, students are required to pass all of the C-level quizzes; to earn an A or B, students must pass a certain number of A-level quizzes as well. Each genre of quiz may be attempted several times during the semester (usually one each Friday) until the student passes the quiz. A similar C-level/A-level distinction is made in assigning homework problems.

The quizzes in Burton's general physics course (and many conventional exams in other physics courses) are produced by a quiz-maker computer program, which automatically inserts randomly selected numbers in key places in each problem and computes the key. Thus, multiple versions of each exam may be produced from one master file, either for use with the same class or for variations from one year to the next. Some of the quiz-maker files have also been transcribed into computer-assisted instruction (CAI) lessons, which the students may use for practice. The CAI system keeps records of each student's progress in the lesson.

Grades for all of Burton's courses are maintained by a computer program that allows students to view their own grades in the course at any time during the semester by way of the campus computer network. The final course grades are also posted by way of the same system. A special program for the self-paced testing system was developed, along with a general gradebook system, for use with any grading scheme based on a weighted average of several types of scores.

Students generally favor the extended opportunities for passing the quizzes, according to Burton. Some complain that the work required for A or B grades is excessive. One of Burton's colleagues in the physics department at Carson Newman uses a similar system.

The software for the quiz-maker, CAI, and gradebook systems was developed by Burton on a digital VAX computer and is written in VAX BASIC. Requests for information will be taken by Burton, particularly for the physics grade program, which can be distributed gratis. The CAI system and general gradebook program are copyrighted.

MULTIPLE-CHOICE PRACTICAL EXAM

Robert C. Capen, Canyon Del Oro High School, 25 West Calle Concordia, Tucson, AZ 85737; TEL: (520) 825-9108; FAX: (520) 292-5927.

Courses Taught

- Physics (high school)

Description of Examination Innovation

In his high school physics and physical sciences courses, Robert Capen creates a multiple-choice "practical" exam in which students move from station to station with a 2-minute task to accomplish at each. Most of the stations have manipulatives that students have seen before in his class; others have manipulatives new to students. Once they are familiar with the manipulatives, the students' challenge is to use the objects in a different reference frame from what they did in class. For example, they use syringes to measure the volume of fluids in class. On the test, they are given similar syringes with the end sealed and the plunger fully drawn to the maximum volume calibration mark. They are asked to compress the syringe and make a qualitative choice among five graphs of force versus compression distance. In the case of manipulatives they have not seen before, students follow an explicit set of directions that allows them to observe a phenomenon they should recognize from class.

Capen has used this test strategy over the past 8 years and reports that it is one of the better ones he has ever constructed for individual assessment.

In another type of test, Capen asks cooperative groups to meet a "lab challenge." The instructor identifies one variable students are allowed to manipulate, and students must develop a model so that they can control all other salient variables. For example, the instructor might provide students with a stopwatch, a steel ball bearing, and an adjustable 6-foot inclined plane. They are to roll the ball down the ramp and predict where its initial point of impact will be on the floor. They may place the lower end of the ramp at the edge of the table or some distance before its end. Instructions require that the incline of the ramp not exceed 35 degrees or be less than 15 degrees from the horizontal.

The point of the exercise is to have the group spend two class periods to develop a theoretical model and test it in lab. To determine who in the group will do the oral explaining once the model is determined, the instructor rolls dice.

After the interview, the instructor gives the group the position on the ramp from which they will roll the ball. They are allowed 4 minutes to perform the calculations and identify the position on the floor where the first impact point of the ball will occur. Their success at doing this will determine the combined grade for the group.

Capen uses cooperative groups for other class projects, often conducting group interviews to ascertain individual learning. He asks one or two questions of each person in a group, starting with a global and often difficult concept or analytical situation, then modifying the question to share a little information with the student.

"Essentially, the instructor can descend the Bloom's Taxonomy scale to determine if the student has competent knowledge, should this be necessary," says Capen. Instructors may weight this aspect of the project in whatever way they choose, says Capen, so long as it is explained to students in advance.

Capen's high school gives teachers a good deal of latitude in exams, but many others have strict curriculum guides in place.

OPTION OF MULTIPLE-CHOICE OR ESSAY EXAM

Stan Celestian, Department of Physics, Glendale Community College, 6000 Olive, Glendale, AZ 85302-3090; TEL: (602) 435-3681; FAX: (602) 435-3329; E-MAIL: celestian@ge.maricopa.edu.

Courses Taught

- Introductory Astronomy
- Introductory Geology
- Physical Science

Description of Examination Innovation

In Stan Celestian's physical science course, students have the option of taking a traditional multiple-choice test or an essay test, or both, with the higher grade of the two recorded. To encourage students to try the essay format, Celestian distributes the essay questions to students 1 week before the multiple-choice tests. Typically, there are 20 essay questions handed out in advance, from which the instructor picks 11, from which the student picks 10. The questions also serve as a study guide for the other students.

If students are not satisfied with their multiple-choice test grades, they can opt for a "makeup" essay test so long as no more than 1 week has elapsed. (There is no multiple-choice "makeup" for the essay exam, because the instructor allows students to keep their question sheets.) If a student misses a multiple-choice test for any reason, his/her only makeup option is an essay test.

In teaching students how to take essay exams, Celestian emphasizes "completeness" and "college-level" answers, which he outlines in class. The essays are not expected to be long: Up to two or three paragraphs suffice. He also advises students not to use the essay questions *exclusively* as a study guide for the multiple-choice test because, although the same material is covered, certain types of information lend themselves to each of these formats. Some students elect not to take the multiple-choice test and then do poorly on the essay exam. But this is their choice: Celestian says students understand the grading system and do not quibble.

The time required to make up and grade the extra tests may discourage colleagues from adopting the practice, but Celestian has been employing this grading system with much success for the past 6 years.

GROUP EXIT INTERVIEW SUBSTITUTED FOR LAB REPORT

Lawrence Day, Department of Physics, Utica College of Syracuse University, 1600 Burrstone Road, Utica, NY 13502; TEL: (315) 792-3099; FAX: (315) 792-3292; E-MAIL: lhday@mailbox.syr.edu.

Courses Taught

- General Physics II: Electricity, Optics, and Modern Physics

Description of Examination Innovation

As part of his attempt to update the laboratory section of physics 152, the second semester of an algebra-based introductory physics course, Lawrence Day experimented in the spring semester of 1993 with the substitution of an exit interview instead of the usual physics lab report for certain of the labs. For these labs (about six per semester), students were required to meet in their four-person lab group with the instructor at the end of the lab period.

During the exit interview, which Day continues to use, students are asked individually about the important ideas contained in the lab, for example, "What was the lab really about?" "Why do you think the lab was structured the way it is?" "What, exactly, were you getting at in the second section of part iii?" and "Do you think the lab was trying to teach you anything new, or was it re-enforcing ideas that you already had from lecture?" In addition, the instructor asks for feedback (e.g., "In your eyes was the lab a success?" "What changes would you like to see in the lab?") and asks about the group process: "What part did you play in your group's completion of the lab?" "Why did your group decide that each of you should do his/her own measurement in part iv?" And finally, he asks whether results were as expected, and if so, why, and if not, why not?

Although he starts off by throwing out a question to anyone who wants to answer, the instructor's follow-up questions are always directed to individual students so he can probe each student's individual understanding. Sometimes, he will ask one student to comment on what was said by another, or to provide an additional answer to a question already answered by a classmate. "This helps with the tremendous difficulty involved in separating group work from individual work when grading lab reports," says Day.

Day thinks he succeeds in getting students to think more deeply about their lab assignment and about their personal contributions to the group's efforts. The students are not as uniformly positive about the method. Better students prefer the exit interviews to lab reports. Weaker students (the instructor thinks) may be regretting the chance to "bluff" their way through in a lab report. Most students agree, however, with the instructor's purpose: that the real learning in a lab course takes place during the lab period, and not in the writing of the lab report. The lesson: never leave the lab room unsure of what was done, why it was done, or how it was done.

Day suspects faculty will not immediately rush to use this technique: "Faculty may always be resistant to giving up the long-cherished lab report. After all, we've all grown up with it and regard it as an integral part of any science lab." Still, he hopes colleagues will question their assumptions about the role of the lab in undergraduate science, and if they do, that they will try exit interviews.

DISCUSSION QUESTIONS COMPLETED BY STUDENT PAIRS

Maria C. Di Stefano, Division of Science, Truman State University, Kirksville, MO 63501; TEL: (816) 785-4583; FAX: (816) 785-7604;
E-MAIL: msidtefa@truman.edu.

Courses Taught

- Introductory Physics
- Advanced Laboratory Physics

Description of Examination Innovation

Maria Di Stefano has a practical solution to the task of assigning students conceptually challenging questions in her two-semester, algebra-based physics course. "Any textbook problem," she believes, "can be enhanced if students are asked 'why' questions, to explain certain aspects of the problem, or to speculate as to 'what if...' or 'describe in your own words...'"

In an unpublished paper, "Nontraditional Testing Techniques in Introductory Physics Courses,"[1] Di Stefano links her testing innovations to her twin teaching goals: first, for her students to achieve conceptual understanding of physics principles; second, to make it possible for "students with diverse styles" to excel. Her testing innovations are embedded in her new two-semester-long introductory course based on a hands-on approach to the construction of basic concepts. In this course, with between 24 and 72 enrollees, the distinction between lectures and labs has become hard to draw, since lab activities include questions and short exercises to be discussed with partners, demonstrations, and explanations by the instructor, all of which are also found in her lectures.

Recognizing the need to invent new ways of evaluating students, Di Stefano writes in her paper the difficulties involved:

> It is very difficult to design test questions and written assignments that will address higher-order thinking processes and that will probe students' conceptual understanding. Such questions are much more arduous to develop than typical end-of-chapter problems, and one person rapidly exhausts his or her inspiration. In addition, non-numerical responses are more difficult to grade, demanding more time from the instructor, and they also deserve more elaborate and individualized feedback.

In addition to problems she creates, Di Stefano borrows from a new collection of physics problems published by Arnold Arons.[2] In both instances, she has adopted the following principles from which she develops test questions: (1) Problems seem more real if numerical data are obtained experimentally instead of being a collection of numbers given as information; and (2) it is beneficial to expose contradictions in physical situations and ask students to deal with these. Such problems encourage critical thinking, logical reasoning, and integration of knowledge.

One example is this one: Students are asked to draw electric field lines (a standard exercise) and then to answer a series of questions, including some "How do you know?" questions, which force the student to think further about the *meaning* of the field lines. In her "lab final," students must choose from an array of demonstrations and use what they have learned in the course to answer questions about the physics embedded in the demonstration. Di Stefano's goal is to mix calculations and observations and to require written explanations on all her exams.

---

[1] Available from Maria Di Stefano at the address above.

[2] Arnold B. Arons, *Homework and Test Questions for Introductory Physics Teaching* (New York: John Wiley, 1994).

Student reaction is mixed. On the one hand, Di Stefano's students perceive her tests to be difficult, because without a deep understanding of the material, students are less successful on her exams than they were on "plug-and-chug" tests in other settings. Student resistance and the added effort it takes to create and grade these evaluations are downsides to the innovation, but Di Stefano is committed to the new type of test, because her students are making "substantial progress in analytical and logical thinking skills," and because she almost never hears the familiar "I can't do math" or "I have never been good at word problems" complaints from students now. Her next goal is to devise strategies to help her students become aware of this kind of improvement as they proceed through her course.

FOCUSED TRUE–FALSE QUESTIONS

Judy Franz, Department of Physics, University of Alabama–Huntsville, Huntsville, AL 35899; TEL: (301) 209-3270; FAX: (301) 209-0865; E-MAIL: franz@aps.org.

Courses Taught

- Introductory Physics (noncalculus for premed students)

Description of Examination Innovation

Eager to find a way to test understanding of concepts efficiently (in addition to normal problem solving), Judy Franz has tried multiple-choice and true–false questions on examinations in her introductory physics course for premedical students. Her current favorite is a composite: the focused true–false question, which consists of 4–5 true–false statements on the same general situation, and requires students to state whether each statement is true or false.

A typical question states, "An ammeter is made from a galvanometer with a shunt resistor in parallel with it." Students must then indicate true or false for the following four statements:

a. The full-scale reading of the galvanometer is always larger than the current through the ammeter.

b. The shunt resistor is usually much larger than the resistance of the galvanometer itself.

c. The current must go through the pointer on the meter for the ammeter to work.

d. The galvanometer works because there is a force on a current-carrying wire in a magnetic field.

The instructor finds she can include about 5 of these focused true–false questions (which incorporate 20–25 gradable answers by the student) on a single test and still take considerably less of her students' time than would be required to answer 20–25 individual true–false questions. The reason for this advantage over traditional true–false questions, Franz finds, is that her students need to read less and consider fewer unique situations, allowing more time for thinking about concepts. This allows the instructor also to include 3–4 traditional problems without students being too pressed for time. An advantage over traditional multiple-choice exams is that each true–false question can be sharply focused and needs only count 1–2 points, whereas multiple-choice questions are usually counted 5–7 points, and a wrong answer often fails to indicate which concept is misunderstood.

Franz admits that this is not a "radical change," but she finds it effective. Some students love the questions, others hate them, but they all say that they have to think to answer them, and Franz likes that. It is interesting, she says, that some students who do well on traditional problems have real trouble with these questions, whereas others who have trouble with traditional problems do fairly well on her focused true–false questions (some, of course, excel on both, and some on neither).

PICTORIAL EXAM QUESTIONS

Shila Garg, Department of Physics, The College of Wooster, Wooster, OH 44691; TEL: (330) 263-2586; FAX: (330) 263-2516;
E-MAIL: sgarg@acs.wooster.edu.

Courses Taught

- Calculus- and algebra-based general physics
- Discovery-based introductory labs
- Physics for Non-Scientists
- Astronomy
- Modern Optics
- Electronics (with lab)
- Modern Physics
- Thermal Physics
- Mechanics
- Advanced Labs
- First-Year Seminar
- Sophomore Seminar

Description of Examination Innovation

Shila Garg has been experimenting with a new exam format in her course for nonscience majors in physics, namely, the addition of pictorial questions that require students to make observations and then discuss, discriminate between, or calculate the results of the illustrated phenomena. For example,

1 Show how light will be reflected from a rough surface by sketching reflected rays for 1 and 2:

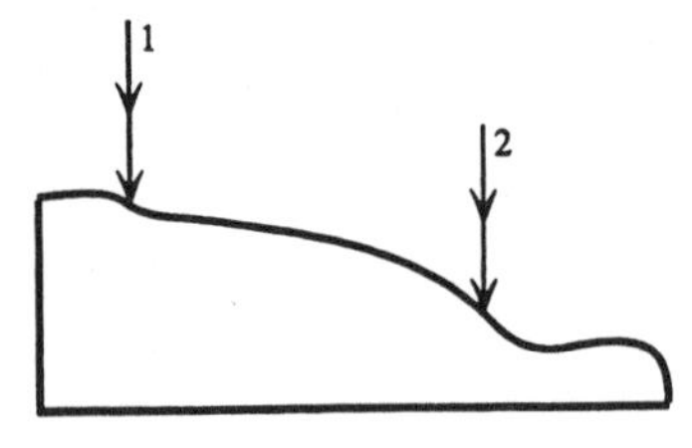

2 In figure (c) below, what is the acceleration?

Figure a
A net force accelerates the crate a α F/m

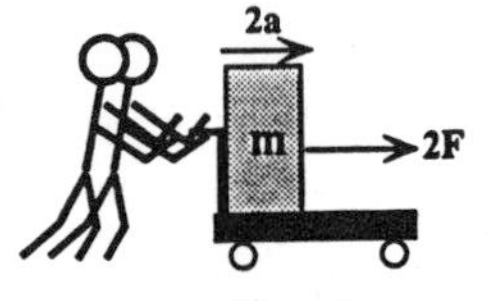

Figure b
If the net force is doubled, the acceleration is doubled.

Figure c
If the mass is doubled, the acceleration is ?

3 As one moves away from Earth, the force of attraction due to gravity decreases. Write down the force attraction at point D.

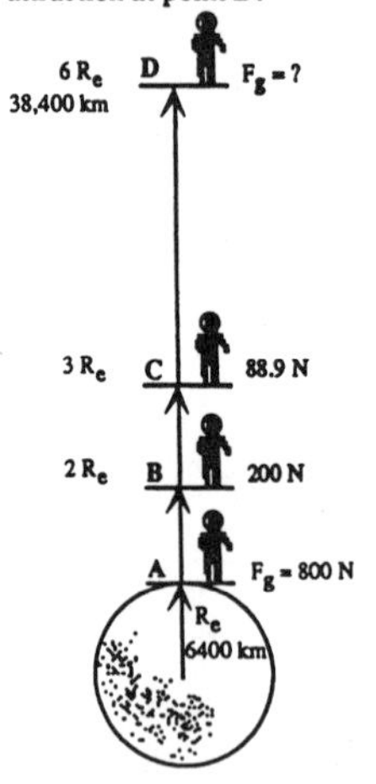

4 Assuming that scales on (a) and (b) are the same, which of the two waves has

(i) greater frequency?

(ii) greater amplitude?

(iii) greater wavelength?

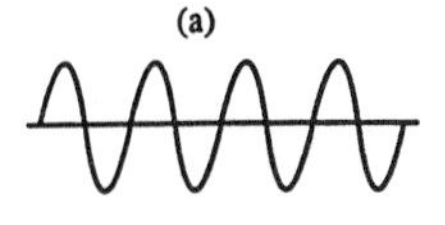

5 Why does the bomb end up where it does rather than at point A?

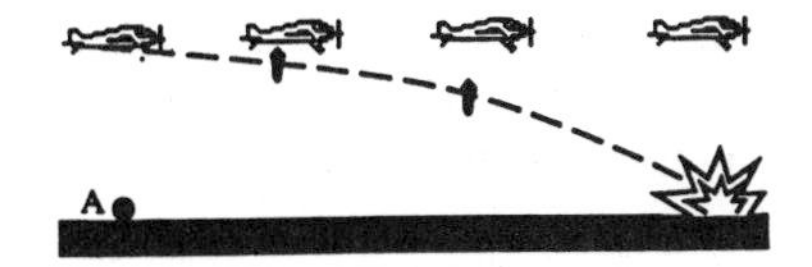

The effort grew out of her observation that multiple-choice questions are confining and difficult for the kind of student who opts for the nonscience option. The mathematical and/or abstract language commonly used to communicate physics also intimidates nonscience majors. In her teaching, Garg experimented with other media, such as audiovisuals and overheads, and found them more effective at keeping nonscience majors' attention and stimulating their interest. She then started including pictorial questions in quizzes and tests.

Garg's class average has improved since she began using pictorial questions. She has also found a remarkable improvement in her students' ability to communicate their understanding of physics. Class evaluations indicate students prefer this style of questioning.

Garg makes up most of these questions herself. She recommends the text *Physics: A Practical and Conceptual Approach*[3] as having good pictorial examples and exercises.

---

[3] Jerry D. Wilson, *Physics: A Practical and Conceptual Approach* (Philadelphia: Saunders, 1993).

MODIFIED MASTERY SYSTEM

Elsa Glover, Department of Physics, Stillman College, Box 1430, Tuscaloosa, AL 35403; TEL: (205) 366-8948; FAX: (205) 366-8996.

Courses Taught

- Elementary Physics
- Intermediate Physics
- Advanced Mechanics
- Advanced Electricity and Magnetism
- Optics
- Thermodynamics
- Modern Physics
- Physics Seminar

Description of Examination Innovation

Elsa Glover has been using a modified mastery system for exams for over 27 years. During most of that time, students in all her physics classes, if dissatisfied with their grade on the first exam, could take different makeup exams as many times as they needed until satisfied. Students kept their highest score without any penalties for makeups. Students really appreciated this technique, and if she ever forgot to schedule a makeup exam, they would request time to take it.

Last year she extended her approach to require students to achieve mastery on each topic in the elementary physics class before receiving credit for it. She divided her elementary physics class, with between 30–50 students, into approximately 15 week-long units such as "unit conversion," "representation of vectors" and "conservation of energy." Each unit includes one lecture, one problem session, and one mostly quantitative exam. Students take the exams and receive pass/try-again grades (a pass grade requires demonstration of mastery, although small calculational mistakes are not penalized), and they must keep on taking each unit exam until they pass it. There is no prescribed order for a particular test. Students need not master one unit before going on to another. Every Friday, all the tests on all the units covered to date are available.

This method gives everyone an opportunity to try for an A. It is particularly helpful for students who have had difficulty performing well in science in the past. "It's a little like a video game as students keep playing until they win," says Glover. Test anxiety decreases as well with the knowledge that one can try again.

The drawback of the method is the vast number of tests that must be created and graded. Although unit tests are short "single-idea" exams (one 15-minute problem with three or four parts), the instructor creates about 20 versions of each.

"Building the files initially is where the additional work comes in," explains Glover. Once the exams are created, however, the instructor doesn't redo them every year. With so many versions for each unit, Glover isn't concerned with students sharing them during a semester or from one year to the next. Students never know until Friday which of the many versions will be given. And the instructor takes the position that if a student can work any of the 20 test problems, he/she has probably mastered the unit.

The exams are short enough that a student who has fallen behind can take not only the current unit test but also several back tests as well.

Glover decided to explore mastery learning further after reading William Glasser's *The Quality School.*[4] Ideally, she said, students wouldn't get credit for the course until they'd mastered all the unit exams. However, to accommodate student grading requirements, she uses the number of exams passed to determine final course grades. In her second semester of the course, she does allow students to continue taking first-semester exams for the first month and raises their first-semester grades accordingly. Also, students who reregister for either semester may take up where they left off, and thus may continue until they get an A.

In her previous method, students didn't necessarily achieve mastery. There, they would be given points for what they got correct, and "passing" was calculated at 55 percent. Now, students do not get any credit (do not pass) until they demonstrate mastery by getting a requisite number of problems correct. Students' attitudes change, says Glover, when they know mastery will be required of them. "Teachers moan about students' sloppy work, but they need to see that students *can't* do quality work unless given the time and opportunity to keep at things until they master them," she adds.

A number of teachers at Stillman use some form of exam retakes, but none, as yet, has gone as far as Glover with the mastery approach.

INQUIRY-DRIVEN QUESTIONS BASED ON PERSONALLY AND SOCIALLY RELEVANT SCIENCE TOPICS

Robert M. Hazen, Department of Physics, George Mason University, 4400 University Drive, Fairfax, VA 22030; TEL: (202) 686-2410 Ext. 2470; FAX: (202) 686-2419; E-MAIL: hazen@GL.ciw.edu.

Courses Taught

- Great Ideas in Science
- Image of the Scientist in Popular Culture
- Symmetry in Art and Science

---

[4] William Glasser, *The Quality School: Managing Students without Coercion* (New York: Perennial Library, 1990).

- Scientific Ethics
- Visual Thinking

Description of Examination Innovation

In an unusual multidisciplinary science course for nonscience majors at George Mason University, from which the textbook, *The Sciences: An Integrated Approach*,[5] was created, Robert Hazen and Jim Trefil are experimenting as well with exam content and formats. In the course, called "Great Ideas in Science," the first exam consists of playing a sports video backwards and asking students to write a short essay on what natural laws are violated, which is another way of asking: Why did you laugh?

Elsewhere in the course, current newspaper articles are employed in examination questions, requiring students to demonstrate their ability to apply scientific concepts (e.g., DNA testing, genetic diseases, nuclear waste) to news events.

Most exams are in essay or short-answer format. In their textbook, Trefil and Hazen provide both "investigation" and "discussion" questions, indicating that instructors should require students to do both and use similar sorts of inquiry-driven questions on exams. "Discussion" questions ask the student to develop a concept or opinion based on material in the text. For example, in the material on thermodynamics, they ask: "Identify three examples of the second law of thermodynamics in action that have occurred since you woke up this morning." "Investigation" questions require additional outside research: "What kind of insulation is installed in your home? What could you do to improve your home's insulation?"

CONCEPTUAL WRITING EXERCISES, ESSAY QUESTIONS, GROUP EXERCISES

C. S. Kalman, Department of Physics, Concordia University, Montreal, Quebec, Canada H3G 1M8; TEL: (514) 848-3284; FAX: (514) 848-2828;
E-MAIL: kalman@vax2.concordia.ca.

Courses Taught

- Introductory Mechanics
- Special Relativity
- Classical Electrodynamics
- Theoretical Physics

[5] James Trefil and Robert Hazen, *The Sciences: An Integrated Approach* (New York: John Wiley, 1994).

Description of Examination Innovation

Despite the large enrollment in his introductory mechanics course at Concordia University, C. S. Kalman uses a student-centered learning approach to make students aware of the concepts underlying the topics being discussed. Journals, use of a tight course outline, and small cooperative group work in class form the basis of Kalman's system.

Each week, Kalman requires students to read and respond to the assigned material for next week's lecture in their journals. They write freely about what they've read, filling at least three pages (there's no upper limit). After rereading their work, students are required to pen three sentences about three important concepts they learned, to be discussed in class that week. All the writing, plus the three sentences, are handed in at the beginning of the class.

Students are graded on the three sentences only, but they must turn in the freewriting to receive a grade (each student's nine best concept assignments are worth 20 percent of the total course grade). Kalman says the freewriting is based on the "writing to learn" concept: Writing that flows without structure engages the whole mind on the task at hand.

"I was initially skeptical," says Kalman, "but then I discovered C-level students asking probing questions and answering their own questions in their freewriting at the level of A students."

During the lecture, the instructor assumes that students have read the material in advance. Having students familiar with the topic and being able to leave out material they read on their own allows him time for in-class cooperative group work, demonstrations, and discussion of problem-solving methods—activities he claims he wouldn't have time for otherwise. Students work in class on "conceptual learning exercises" in groups of four or five, identifying, for example, the forces acting on a thrown baseball. Each group is supplied with a transparency and a marker, and a designated "reporter" records group members' responses to the problem on the transparency. At the end of the period, two groups with differing viewpoints are chosen to display their transparency to the class and report on their results. Once the differing views are established, the role of experiment in deciding the issue can be emphasized. A demonstration can then decide the issue.

These conceptual learning exercises, and the writing requirements throughout the course, foreshadow essay questions used on the midterm and final exams. One question (minimally) covers the topic done in group work during class—the action of forces on a projectile such as a tennis ball struck by a tennis racket. In Kalman's class, almost all the students who participated in the group discussions correctly answered this question.

In his more advanced special relativity course (with fewer students), Kalman requires one-page miniessays every week summarizing and interpreting the weekly lectures. This past year, he expanded this requirement to include creation of a course "dossier," a collection of eight or more short, one-page summaries of what was covered in class, unified into a single overview based on common themes.

"I have been doing various forms of essay grading in physics for many years now," says Kalman, "and not one student has protested that the grading is 'too subjective.' Many say they wish all professors would use this technique, and many continue to use journalizing in other courses."

Group work is also a part of special relativity. Every 2 weeks there, is a group exercise on philosophies of science, during which groups debate different epistemological points of view. Each group takes a philosophical position (for purposes of debate) for the entire term, and the philosophical debates find their way onto the final examination.

Kalman credits his wife, Judith Kalman, who has had many successes in teaching writing at Concordia University, with inspiring much of his effort to bring writing into the science classroom.

COLLABORATIVE LEARNING, TEAM EXAMS AND PROJECTS, ACTIVE LEARNING PROBLEM SHEETS, MBL PROBLEM-SOLVING, CLASSROOM ASSESSMENT

Oshri Karmon, Department of Physics, Diablo Valley College, Pleasant Hill, CA 94523; TEL: (510) 685-1230 EXT. 528; FAX: (510) 685-1551;
E-MAIL: okarmon@dvc.edu.

Courses Taught

- Elementary Physics
- Introductory Physics for Engineers and Scientists

Description of Examination Innovation

Among the testing innovations Oshri Karmon has been using for the past 5 years are team exams linked to extensive class work in groups. After a minilecture (about 10 minutes) by the instructor, pairs of students work on group exercises. These assignments start with simple calculations and progress to more involved contextual problems. Sometimes group exercises involve laboratory-type assignments that require taking measurements to verify concepts.

Each pair is responsible for submitting answers, to be graded, but the pair works at a table together with three or four other pairs. The instructor circulates among the tables while the work is being completed. If a whole table is having difficulty, he sends over pairs from other tables as "consultants." Classwork is collected and graded.

Karmon also allows students to complete collaborative, optional projects for extra credit or to replace failed exams. These lab projects are based on more difficult problems or experiments and are designed by the students with the instructor's guidance. In one project, students predicted and measured the terminal velocity of a ball bearing falling through a viscous fluid. In another, they developed a computer prediction and verified by experiment the frequency of large-angle oscillations of a pendulum.

Karmon uses active learning problem sheets (ALPS) advanced problem solving sets by Van Heuvelen[6] and microcomputer-based labs (MBL) by Thornton and Laws in his physics for engineers class. Finally, he employs classroom assessments such as the "1-minute paper" (described in *Classroom Assessment Techniques* by Angelo and Cross),[7] in which students list the clearest and the muddiest points in a lecture.

Karmon's grading plan has several components—essays, individual and collaborative exams, class work and quizzes, homework, lab reports, and bonus assignments—allowing him to evaluate student performance at different levels. Critical to his plan is personal contact and familiarity with each student, as well as ongoing evaluation to keep the students informed about their progress. When students have difficulty in the class, Karmon alerts them and designs with each a plan to improve his/her standing.

In "town meetings," Karmon and students discuss the pros and cons of collaborative learning. Written evaluations indicate that students feel less anxious about math and science and more interested and confident in a cooperative, rather than a competitive, classroom environment. Karmon says he was drawn to these innovative methods because he himself had flunked physics in high school. He had to retake his exams, and that is when he discovered he could study physics and calculus on his own.

"I realized that anxiety and fear interfere with learning, and that students really need me only as a coach," he says. Now he gives them the tools to solve realistic problems and the responsibility of participating actively in his classes and in the grading process.

INDIVIDUALIZED, COMPUTER-ASSISTED, PERSONALIZED ASSIGNMENT SYSTEM

Edwin Kashy, Department of Physics and Astronomy, and David Morrissey, Department of Chemistry, Michigan State University, East Lansing, MI 48824; TEL: (517) 333-6318 (Kashy), (517) 333-6321 (Morrissey); FAX: (517) 353-5967; E-MAIL: kashy@nscl.msu.edu *or* morrissey@nscl.msu.edu.

Courses Taught

- Physics of the World Around Us
- Introductory Physics I and II (noncalculus)
- Physics for Scientists and Engineers I and II (calculus-based)

---

[6] ALPS problem sets can be requested from Alan Van Heuvelen, Physics Education, Ohio State University, Columbus, OH 43210.

[7] Thomas A. Angelo and K. Patricia Cross, *Classroom Assessment Techniques* (San Francisco: Jossey-Bass Publishers, 1993).

- Introductory Chemistry I and II
- Calculus
- Biochemistry

Description of Examination Innovation

Edwin Kashy and David Morrissey are using an integrated Computer-Assisted Personalized Approach (CAPA) system to create individual assignments, quizzes, and exams for students.[8] CAPA allows students to enter their answers to personalized assignments, which include both quantitative and qualitative questions, directly via networked terminals at the school or from their own computers at home. The system gives student immediate feedback and hints while providing the instructor with on-line performance information. The instructors report that because of the availability of feedback and hints, they have included more challenging questions in homework assignments, which, in turn, has made it possible to ask more challenging questions on exams.

As the system is currently employed for assignments by the Michigan State University (MSU) instructors, students are allowed to reenter solutions to the problems any time before the due date without penalty. This provides the opportunity and an incentive for students of different skill levels to learn the material. The system also records the pattern of data entry, including errors, thus providing valuable information to instructors. Because the problems are unique, students are encouraged to study together, yet each must do his/her own work to obtain correct answers. Since instituting the system in the fall 1992, and at the request of students, explanations of problems have been available to students immediately following the due date of an assignment.

CAPA has been used for the past few years to generate individual midterm examinations in introductory physics classes. Most problems are numerical, with an answer format that requires choosing 1 of 8 numerical answers. Approximately one-fourth of the questions are conceptual, requiring the selection of 2 or 3 correct statements from a list of 5–7 choices. Students can thus do their exams in close proximity, an important issue in large classes, where space between seats is scarce. Students keep their exam papers, and turn in only an answer sheet indicating the choice(s) selected for each problem. They can get feedback immediately following the exam, because they can log in and check their performance.

Instructors who believe that midterm examinations are an integral and important part of the learning experience have also given students the following option: When leaving the examination room, students may pick up a different

---

[8] See *American Journal of Physics* 61, no. 12(1993): 1124–1130, for a detailed description of CAPA by the innovators.

individual version of their exam in which the numerical problems do not have a multiple-choice format. They can they rework any of the problems that they have missed and earn partial credit (typically 30 percent) for corrected questions. Practically all students take advantage of this option for the midterms.

Students appreciate the mastery-type approach, and their instructors believe it generates many more student hours of concentrated effort to learn the material on the exams, which might otherwise have not taken place. Student response to CAPA has been positive. In a student survey of a University of Rochester physics class using CAPA, students said the system is useful and "friendly," and that it encourages confidence and pleasure in problem solving. "It is remarkable," says Kashy, "how many students refuse to stop at 15 out of 16 correct, or 16 out of 17. They appear driven to get that last point. Log-in records confirm that this is a widespread effect." Due to the many hours saved on grading, instructors and TAs have more time for increased office hours and individual attention to students.

Kashy and Morrissey have made CAPA available to other campuses. Non-MSU users can be licensed to use the software by contacting the Office of Computing and Technology or by writing to either instructor using electronic mail. To determine CAPA's usefulness for either assignments or examinations in other disciplines, problems have been coded by potential users in fields other than physics, chemistry, calculus, and biochemistry. These have included botany, statistics, quantum mechanics, and history.

MODIFIED PSI/MASTERY LEARNING SYSTEM

Stephen V. Letcher, Department of Physics, University of Rhode Island, Kingston, RI 02881-0817; TEL: (401) 874-2633; FAX: (401) 874-2380; E-MAIL: sletcher@uriacc.uri.edu.

Courses Taught

- Introductory Physics

Description of Examination Innovation

Based on his experiments with mastery learning examination procedures in introductory physics,[9] Stephen Letcher employs a sequence of "unit exams" in his introductory physics course at the University of Rhode Island. The basic

[9] Described in Stephen V. Letcher, "Mastery-Learning Examination Procedure for Introductory Physics," *The Physics Teacher* (November 1989): 613–614.

premise of mastery-learning exam procedures, according to Letcher, is that "students should not be allowed to do poorly on an exam and then just go on to the next topic."

Letcher's course is divided into four units, and each unit culminates in an hour exam. "It is expected," he tells his students in their course handout, "that you will be well prepared for each unit exam and will do well." But if for some reason this is not the case, students are expected to stick with that unit exam until they get a satisfactory result.

The rules are as follows: If the grade is less than 75 percent the first time the unit exam is taken, then the student must take it again. The second written exam will have different questions from the first but be similar in content and style. If the grade on the second exam is 75 percent or higher, then the grade for the entire unit will be an average of the two tries weighted in the direction of the first: 60/40.

If a student fails to get a 75 percent on the second try, then he/she must see the instructor and make an appointment to take a 20-minute oral exam. In this case, the unit grade is the average of the two written and one oral exams, weighted as follows: 48 32 20. (Students who pass the first exam with 75 percent but want to try raising their grade may also take the second exam under the same rules.)

Letcher says about 80 percent of his students react favorably to this method. In an early evaluation, a student wrote that the new approach "gives an opportunity for a better grade and also, since you are repeating a test in the same unit, you understand it better." Writes another: "I've taken this course before using the traditional hour exam system. I would say I have learned five times more using [the new] system."

The value of this system, from a testing point of view, according to Letcher, is "that it allows the instructor simultaneously to give meaningful test questions, maintain high standards, avoid scaling, and avoid failing untoward numbers of students."

"REDEMPTION" POINTS, MOTIVATIONAL QUIZZES

Robert MacQueen, Department of Physics, Rhodes College, 2000 North Parkway, Memphis, TN 38112; TEL: (901) 726-3915; FAX: (901) 726-3565; E-MAIL: rmac@rhodes.edu.

Courses Taught

- Astronomy
- Introductory Physics
- Dynamics

Description of Examination Innovation

In recent years, in his introductory physics course in astronomy, Bob MacQueen has taken to using "redemption points" on the final exam. This practice reduces students' anxieties and encourages extra effort from students to understand the basic physics material essential to understanding the course. An inverse valuation of the first test score is applied to the appropriate section of the final exam. Thus, a student who scored "50" on the first test can earn 1/0.5 = 2 points for each point on the final.

Because MacQueen wishes to stress to students that they should continue to review and seek understanding of these concepts, redemption points are applied *only* to the first exam of the semester. This scheme is designed to offer a tangible reward for review and mastery of the background physics material. It also offers hope (and support for staying in the class) for the student who does poorly on the first exam.

To encourage daily preparation and class attendance (required attendance is at the option of the instructor at Rhodes) in introductory astronomy, MacQueen offers the following "sporting" quizzes three to four times per semester. He calls on 10 students, randomly preselected, and asks each a different question. If 7 or more of the 10 answer correctly, the *entire class* receives a 1-point addition to their final grade average. If fewer than 7 answers are correct, no extra credit is earned. Absent students called upon each count as a "miss," thus encouraging peer pressure to attend class.

MacQueen's selection mode employs a random-number generator scaled to the number of students in the class. It is possible, and it has happened (indicating the true randomness of the system), that the same student would be called upon twice in one session.

It is difficult to assess changes in class preparation and attendance with and without the quizzes, but the atmosphere on the day of the quiz is suspenseful and enlivening. "Perhaps it is the generation of positive feelings about the class and the material that are the most important aspects of the innovation," suggests MacQueen.

The instructor says he hasn't encountered any drawbacks to using redemption points. Regarding the "sporting" quizzes, however, there are three: First, if the first four answers are incorrect, the suspense is gone and apprehension sets in (although, invariably, students ask that the procedure continue so they can see how close they come); second, substantial peer pressure is exerted on the student faced with the "make or break" question for the entire class. If that student fails to respond correctly, causing the entire class to lose the credit, the student is deeply embarrassed; and third, it is not always so easy to come up with nontrivial but reasonable questions to ask orally.

CONCEPTUAL QUESTIONS, GRADING EXERCISE

Eric Mazur, Department of Applied Physics, Harvard University, Cambridge, MA 02138-2901; TEL: (617) 495-8729; FAX: (617) 495-9837; E-MAIL: mazur@physics.harvard.edu; PEER INSTRUCTION WEB SITE: http://mazur-www-harvard.edu.

Courses Taught

- Introductory Physics (Physics 11)

Description of Examination Innovation

Five years ago, after changing his teaching of Physics 11 to focus more on conceptual understanding, Eric Mazur began altering his in-class examinations as well by adding more conceptual questions to the final exam[10] Typically, approximately half of the questions on an examination are qualitative and do not require much, if any, calculation. Students are advised to *describe* their answers and to briefly justify their reasoning. A typical example of a conceptual examination question is the following:

> Two carts, A and B, travel toward each other and collide. You may ignore the effects of friction. After the collision, the two carts have the same relative speed, but cart A has lost some kinetic energy.
>
> a. What can you say about the kinetic energy of cart B?
> b. Would the changes in the carts' kinetic energies be different if you put a spring between the two (ignore dissipation in the sprint)? Explain.
> c. And if you put a piece of Celco between them? Explain.
> d. In which of the above situations (a: original, b: with spring, c: with Velcro) is the total momentum conserved? Why (or why not)?

The conceptual exam questions are similar to the multiple-choice conceptual questions, or ConcepTests, which Mazur asks during class meetings to push students to think about the course material; the main difference is that on the exam, the questions are posed in an open-ended fashion rather than in a multiple-choice format. These ConcepTests are available for anyone to use via his/her Web site and have also been published in *Peer Instruction: A User's Manual* (see footnote 10). Many instructors at schools ranging from local branches of state universities to private colleges and universitites have successfully used ConcepTests in their teaching. Students have responded very positively to the use of

---

[10] See Eric Mazur, *Peer Instruction: A User's Manual* (Upper Saddle River, NJ: Prentice-Hall, 1996) for a detailed description of the Peer Instruction method of teaching, including a complete set of ConceptTests and a representative sampling of conceptual examination questions.

these conceptual questions in class; they participate enthusiastically and prepare faithfully for class. (For a more extensive description of the peer instruction approach to classroom teaching, consult the Peer Instruction Web site address given on p. 170 or http://gmlileo.harvard.edu.)

The instructor eschews curved grading (grading is done on an absolute scale), he takes advantage of every opportunity to educate his students about test-taking and grading practice. For example, the course syllabus distributed to students on the first day of class provides a "rough overview" of the grade statistics for Physics 11 over the past several years. Typically, 45% of enrolled students receive As in the first semester (30% in the second semester) and between 2–5% fail the class.

Another method Mazur uses to teach students about testing and test taking is to have them compare their grading with that of their teaching fellows. For this purpose, early in the course, Mazur offers students a "grading exercise" made up of four representative student answers to single questions that appeared on the last 1-hour exam. He distributes these answers among the students and his teaching assistants and has them grade and, more importantly, comment on the answers. The grading scale used is a 4-point scale modeled after the recommendations used in peer review of scientific manuscripts [3 = acceptance, 2 = acceptance after minor revision, 1 = major work needed (after which a second review is possible), and 0 = rejection]. Both students and teaching assistants grade the questions, and then all grades are tabulated and compared. Staff and student grading turn out to be statistically indistinguishable. Most importantly, students realize what is needed to make a solution clear for another reader. The grading exercise begins with this rationale:

> A question we've been asked very often lately is: "How much do I need to write down when solving a problem?" The best way to get a better feel for this is to do some grading. For this reason, we've put together four different solutions for problem 3 on the first hour exam. Although written by us, these solutions are representative of what we see.

10-MINUTE GROUP DISCUSSION HALFWAY THROUGH THE EXAM

John McGuire, Science–Mathematics Division, Porterville College, 100 East College Avenue, Porterville, VA 93257; TEL: (209) 781-3130; FAX: (209) 784-4779; E-MAIL: jmcguire@pc.cc.ca.us.

Courses Taught

- Introductory Physics (trigonometry- and calculus-based)

Description of Examination Innovation

In this small rural community college, John McGuire has been able to experiment with in-class testing strategies. To begin with, he employs a 3-hour lab period to give a consolidated multiple-chapter test. Somewhere in the middle of the 3 hours, students are allowed 10 minutes to go to an adjoining room without their test papers and discuss anything they want with other students. There, in the adjoining room, they can write on the chalkboard, use a calculator, and write notes from the book (most of McGuire's tests are open book). But none of this writing can come back into class.

The instructor finds that the break eases stress and anxiety. Less prepared students do *not* "get all the answers" from friends, in McGuire's experience, and the discussions often become exciting and informed debates. McGuire also finds that from student discussions he gets ideas as to how to improve test content. In any case, he reports, "eavesdropping" on student discussions of test items is interesting and an important source of feedback as to their level of understanding. Students find the method more helpful on tests involving problem solving than those requiring only factual information and recall.

Colleagues have a range of responses to these discussions, including interest, surprise, and guarded skepticism. McGuire cannot use the method in large classes, because he cannot find an available extra room. He also hesitates to try these discussions in classes that do not seem to have developed a cooperative spirit, though he admits using the method might generate cooperation.

PRECOURSE QUIZ ON PREREQUISITE MATERIAL, OPTIONAL FINAL EXAM, INDIVIDUALIZED HOMEWORK ASSIGNMENTS

E. S. Oberhofer, Department of Physics, University of North Carolina at Charlotte, Charlotte, NC 28223; TEL: (704) 547-2536; FAX: (704) 547-3160; E-MAIL: eoberhfr@uncc.edu.

Courses Taught

- Introductory Physics I and II

Description of Examination Innovation

Ed Oberhofer has been teaching an introductory course in physics for nonphysics majors in which he has introduced three testing innovations.

The first is a quiz on the first day of class, which has two purposes: first, to alert students to the fact that "most of them do not have good enough recall of things they've learned in the past"; and second, to demonstrate to them that they should not overlook the material in the first chapter—which, on first glance, might appear too basic to require studying. The quiz serves as well to remind students that they will have to relearn units in a dimensional-analysis fashion, and that they must review some basic formulas pertaining to areas and volumes of some regular geometric figures.

The second is a means of differentiating selected homework problems, which are graded, so that few students are assigned the same problem. Oberhofer calls these TIPs, or turn-in-problem sets. Because they are different (the TIP has a variable in it corresponding to a digit in the student's ID number), the instructor can use the grades on these "homework" problems in place of weekly quizzes. This, he reports, gives him more lecture time. TIPs are one or two problems long. Other homework (about 20 problems per week) is assigned but not evaluated or collected.

A third innovation is an optional final exam. A student whose grade is high enough (from his/her point of view) at the end of classes is allowed to accept it as the semester grade. If it is not high enough, the student can opt to take the final exam and replace the lowest test grade in the averaging that will occur. Oberhofer has found this option motivates both better students and those whose achievement is not as high. The better students, he reports, "really like the reward of not taking the final." This gives them more time to study for other finals, and the option makes them work harder all semester for the higher grade. Students who don't start off the course well (about half the class) have an alternative—if they are willing to work hard—to repeating the semester. The optional final also eliminates the tendency to quit the class after an initial disaster on the first test.

Oberhofer notes that more students are persisting in his classes, and that more students are passing and fewer failing after applying these methods. He has also found that since he made the final exam optional, grades on tests tend to go up as the semester progresses, a trend not recognizable previously. Approximately one-third of students still attending at the end of the semester take the final exam, and it is mostly those who had one bad test who benefit greatly from the final. In course evaluations and in comments to Oberhofer, students report valuing the optional final exam.

The instructor uses the same innovations in the second semester of introductory physics, because the students are used to them. (He now teaches almost all the first-semester students.) An interesting result is that the students taking Introductory Physics II from a different faculty member do as well as those in Oberhofer's class (95 percent passing), suggesting that not taking the final exam did them no harm in their preparation for the next course.

GROUP EXAMS

David G. Onn, Department of Physics and Astronomy, University of Delaware, Newark, DE 19716; TEL: (302) 831-2680; FAX: (302) 831-1637; E-MAIL: dgonn@udel.edu.

Courses Taught

- General Physics (for life and health science majors)

Description of Examination Innovation

There's so much noise coming from David Onn's physics classroom during his "collective" exams that faculty next door sometimes complain, but he is not concerned. As he sees it, his students are learning more physics than they ever did before.

Onn began using active, cooperative learning techniques 4 years ago, when the entire university was moving in the direction of problem-based learning. Initially, he employed the new techniques in lecture and discussion sessions, admonishing his TAs never to solve a physics problems for the students, but rather to send them to the board to solve them for one another. Only later did he extend cooperative problem solving to the exams themselves. When he noticed that even with group exams, his students still did their homework alone, he developed the full-fledged group learning system that is now in place.

Onn's physics class is considered "college physics," in that it is algebra- and trigonometry-based, not calculus-based. His 120–150 students exhibit a wide range of backgrounds, capabilities, and levels of interest in physics. What caused him to create some of his more "radical" systems was the challenge of applying cooperative learning to large classes. "All the examples I was reading about took place in small classes of 20 students. I asked myself whether I could carry some of these ideas over into a much larger class, with average, not honors, students."

In its mature version, at the beginning of the term, Onn assigns students into working groups of three based on previous academic records and a suitable meeting time for the semester. During lecture periods, he intersperses formal minilectures, in which new material is presented, with interactive learning activities, frequently breaking the class into their working groups to discuss homework problems, in-class assignments, large-screen interactive software examples, and to take group quizzes. About 25–30 percent of the students' hour exams (Onn gives three and a final each term) are group problems they solve together. In grading the problems, all the group members get the same score.

These group problems are not "textbook" problems, according to Onn. They are conceptual problems based on real-life situations to which students can relate. The group problems, which take approximately 30 minutes to solve, also provide a review of the material. Often, students take the group portion of the exam on a different day.

"I could use these problems as individual items," says Onn, "but they would take the students longer to solve individually, and the students would be more likely to make mistakes."

More importantly, the point of group exams is that the students learn how to learn from each other.

"It's difficult to measure what students gain from these exams," admits Onn, who's been teaching this course for 25 years, "but my feeling is they're learning more—especially the students who start off weaker than the rest. My class averages are higher than they used to be, and my grade distributions have narrowed. I rarely give Ds and Fs now. Yet my top students are doing at least as well as they would have before." Onn uses what he calls a "complicated" cumulative, absolute grading scale that involves self-evaluation and evaluation of other group members in different cooperative learning activities.

Onn's classes are extremely lively, with a good deal of student activity, especially during the group exam. Some groups subdivide the question into parts and work individually; others divide up responsibilities, with one group working the calculations and another doing drawings.

"Generally," says Onn, "they really do work together. That's why it's so noisy during class time. As soon as they are given the group question, students will clamor for space to work in their groups, sometimes taking over my demo bench. They'll do anything they can to get some privacy."

This is in stark contrast to conventional classes in physics. "They're so used to sitting and listening in lecture that about 20 percent never adjust to the activity level I require. The students who like my group exams and group work feel very positive about the method; the majority are neutral. But one reason *I* like the new format is that students now have to attend class. Where I had 50–60 percent attendance before, I now have 90–95 percent."

Onn's final course evaluations ask students if they've learned anything valuable to them other than physics in this course. Many say they've developed greater responsibility for their own learning and that they've improved their cooperative learning skills.

"I've noticed my students are asking more profound, in-depth questions about physics," says Onn. "In fact, a couple times this semester, their questions astounded me. What they seem to be gaining is what I had hoped they would gain: the sense that it's all right to ask questions in science."

RETESTING WITH REFLECTION, ABSOLUTE GRADING, KEYING GRADING TO SKILLS, UNTIMED FINAL

Frederick Reif, Center for Innovations in Learning, Carnegie Mellon University, 5000 Forbes Avenue, Pittsburgh, PA 15213-3890; TEL: (412) 268-5713; FAX: (412) 268-5646; E-MAIL: freif@andrew.cmu.edu.

Courses Taught

- Introductory Physics

Description of Examination Innovation

In his introductory physics course for scientists at Carnegie Mellon University, Frederick Reif introduced a whole series of innovations intended to tie testing to learning and to increase the feedback he provides his students after their exams.

For example, he offers students who get a score of less than 75 percent on any of their weekly in-class quizzes an opportunity to raise their score to 75 percent. If they want to participate in the scheme, they must review their exams and do the following:

1. Identify the mistakes which they made in obtaining wrong answers or solutions—and diagnose the likely reasons that led to these mistakes.
2. Indicate what precautions they will take not to make these mistakes again.
3. Correct their mistakes so as to write out correct solutions.

The purpose of this option is to encourage students to reflect on what they have done and thus to learn from their past deficiencies. Merely allowing students to take the test again, without such prior reflection, is unlikely to be equally effective, Reif believes. Such reflection is designed to contribute to Reif's central goal of teaching students the thought processes needed for physics.

Reif, who has written widely on matters of teaching and learning in physics, is also interested in encouraging students to check their work. To help achieve this aim, he makes a distinction between answers that are wrong and those that are also "nonsensical." (Answers are nonsensical if their deficiencies can be readily detected by performing some simple checks, such as checking the consistency of units, or checking whether answers agree with some simple special cases.) Students are penalized more severely for such nonsensical answers than for answers that are merely wrong.

Reif uses absolute grading throughout the course to avoid meaningless competition among students—especially since he is also trying to use cooperative learning in small student groups. But it seems unwise to focus merely on absolute performance when giving feedback to students, because "students also want to know how they compare with other students," notes Reif.

To increase the useful feedback that TAs can provide in grading, Reif has developed "grading keys" that focus on some specified abilities—for example, the "ability to determine the direction of an acceleration" and "the ability to enumerate all the forces on an object." Each test lists the abilities it is designed to assess. When TAs grade, they can then use a cross-reference to indicate which abilities the student seems not yet to have mastered.

Reif gives students final examinations without a strict time limit. (Although final examinations are officially 3 hours long, Reif lets students spend as long as 4 hours, if they wish.) This has the following advantages:

1. The examination then provides a better assessment of students' understanding. (Suppose that in 3 hours student A can answer five questions and student B can answer only four. But suppose that in 4 hours student

A can still answer only five questions, but student B can answer seven questions. Then Reif believes that student B has a better understanding of the subject matter, although the 3-hour time limit would have indicated otherwise.)

2. Students are less nervous about time pressure and can think better without being impeded by extraneous emotional factors (e.g., a student encountering a question that he/she cannot immediately answer can then more calmly put it aside to be tried later.)
3. It is more desirable that students leave an examination with the feeling that they have done their best, rather than with the (often illusory) impression that they could easily have done better if they had just had 10 more minutes.

Reif's test items are carefully designed to explore deeper understanding and more mature problem-solving strategies. Qualitative questions (e.g., questions asking whether some force on an object is larger or smaller than some other force) are interspersed with quantitative questions. Similarly, he asks students to provide explanatory comments to indicate the decisions they make in solving problems. (Thus, a mere string of equations is not accepted as a satisfactory solution, irrespective of the correctness of the answer.)

Above all, Reif wants his students to approach the world the way scientists do—that is, realize that the aim of science is to make a large number of correct inferences on the basis of a small number of basic principles. (By contrast, many students coming from high school think of science as a large collection of facts and formulas to be memorized—and as numbers to be crunched on their calculators.) Since Reif is trying to impart to his students the kind of *thinking* that is useful in physics, he has to examine his students in such a way as to test this thinking.

In sum, Reif recognizes that to be effective and consistent he has to test what he cares most about.

THE "CONSULT YOUR NEIGHBOR" QUIZ

Donald L. Shirer, Department of Physics, Yale University, 217 Prospect Street, New Haven, CT 06520-8120; TEL: (203) 432-3664; FAX: (203) 432-6175; E-MAIL: donald.shirer@yale.edu.

Courses Taught

- Introductory Physics

Description of Examination Innovation

In an introductory physics-without-calculus course designed for premed students and other nonphysics majors, Donald Shirer and his coinstructor Shiva

Kumar experimented for the first time in 1994 with a modification of the "do-it-yourself" format on quizzes. They allowed students to "consult" with their neighbors on their answers. Many students prefer this quiz format as an anxiety reducer. And although it has occasionally been abused by some (who had their answer sheets entirely filled out by their neighbor), Shirer and Kumar believe that since they've incorporated in-class discussion into the examination mode, participation in in-class discussion overall in the course will increase. They want students to realize that science is consultative, and that they don't have to know everything from memory to succeed in science (and on science exams).

Shirer and Kumar have used this method for two semesters in large lecture classes with approximately 200 students. Quizzes are graded no differently using this method than without it. Most students like the method, according to Shirer, "especially being able to consult with 'smarter' people." Because the quizzes are worth only about 10 percent of the final grade, Shirer thinks the benefits of the method will outweigh the drawbacks. He looks forward to seeing what students say about it in their end-of-course evaluations.

This experiment was modeled on Eric Mazur's "consultant quiz."[11]

ORAL FINAL EXAMS USING NONSCIENCE FACULTY QUESTIONS AND EVALUATED BY INSTRUCTOR

Hugh Siefken, Department of Physics, Greenville College, Greenville, IL 62246; TEL: (618) 664-2800 Ext. 4472; FAX: (618) 664-1373;
E-MAIL: hsiefken@greenville.edu.

Courses Taught

- Energy and the Environment

Description of Examination Innovation

Every third year in his course on energy and the environment, which meets Greenville College's general education laboratory-science requirement, Hugh Siefken replaces the traditional comprehensive, in-class, written final exam with individual 20-minute oral testing given by nonscience faculty but evaluated by the instructor.[12] The duration is generally 2 days.

---

[11] Described in "Harvard Revisited," Sheila Tobias, *Revitalizing Undergraduate Science* (Tucson, AZ: Research Corporation, 1992), pp. 114–122.

[12] In many European countries, notably Italy, *all* science exams are given orally.

Prior to their oral exams, students are given a list of the major topics covered in class. This same list is sent to faculty colleagues, who volunteer to come for an hour to examine Siefken's students on these topics. As course instructor, Siefken gives instructions to faculty colleagues not to get stuck for more than 3–4 minutes on any one topic. While the testing is going on, Siefken sits back and evaluates students' ability regarding the following criteria: (1) ability to apply the pertinent ideas in the course to the question asked; (2) communication skills; and (3) overall poise in handling the "naive" questions faculty colleagues raise. "You can learn a lot and very quickly," he reports, "during a 20-minute session."

Two features emerge from his innovation. First, students prepare at a much higher level for this test than for traditional tests, probably because they do not want to be embarrassed. Second, students learn that they know much more about the course material than outside faculty members who teach outside the sciences. According to Siefken, well-prepared students love this method of testing, while less-prepared students do not.

"I think it is an indicator of the maturity level of the student," he states. Siefken started using this method 16 years ago, because he wanted to include some basic communication skills in the course that would simulate future situations his students would face. "They need to be able to *deliver* what they know in a way that is understandable to a person not acquainted with the facts at hand," he says.

Because it is so time-consuming to schedule and implement this innovation, Siefken cannot use it every year. A few colleagues who participated as interlocutors tell Siefken they would like to use this format in their classes in the future.

PORTFOLIOS

Timothy F. Slater, Department of Physics and Astronomy, Montana State University, Bozeman, MT 59717; TEL: (406) 994-0211; FAX: (406) 994-1789; E-MAIL: tslater@math.montana.edu *or* http://www.math.montana.edu/~tslater/.

Courses Taught

- Physics
- Astronomy
- Geology

Description of Examination Innovation

Tim Slater teaches physics from a constructivist perspective, and so he tries to incorporate this philosophy into his assessment strategy. Capitalizing on his observation that students only learn what they have to learn to achieve the grade they desire, he decided in 1993 to provide his students with a specific set of guidelines and objectives in advance of their tests. He selected what he decided were the 24 most important objectives for first-semester college physics and asked his students to be prepared to "document their adventure in learning physics."[13]

Objectives included the following:

- Understand the nature of scientific knowledge and the various disciplines of science.
- Appropriately apply vectors qualitatively to describe physical situations.
- Use vectors to quantitatively solve problems relating to motion.
- Solve problems related to static equilibrium and rotational equilibrium.
- Apply the Doppler Effect to physical situations quantitatively and qualitatively.

Students document mastery of these objectives through the creation of a personal journal (private) and a formal assessment portfolio (public). Students can place anything in the portfolio that clearly demonstrates mastery of a specified learning objective. Each item submitted as evidence must be clearly labeled, state why the evidence demonstrates mastery, and include statements of self-reflection. Several portfolios have contained essays concerning news clippings that demonstrate a scientific concept; journal entries that describe how a topic is relevant and important to the student personally; homework assignments suggested in class; short research papers; old tests from other classes; study guide quizzes; excerpts from an in-depth class discussion among peers; relevant laboratory reports; reactions to the material presented in the text; and creative homework solutions. The instructor found these creative homework problems and solutions entertaining to grade and revealing of the comprehension of the student.

In his directions to students, Slater makes it clear that the portfolio is "your responsibility to demonstrate your knowledge to the instructor." He also requires self-reflection, which he describes as "some indication as to how well you think you understand this objective, how this objective might be useful to you in the future, and how you went about mastering the concept." Since the portfolio constitutes 60 percent of the final course grade, students are encouraged to devote considerable time to it.

Slater's purpose is obvious: to give students credit for active and self-directed learning rather than for excellence in performance on a short-duration, graded task. In comparison studies of students examined in traditional tests with those in his classes, he find that his students perform just as well on traditional exams as their peers, and that they feel they *know* the material better.

---

[13] See Timothy Slater, "Portfolios for Learning and Assessment in Physics," *The Physics Teacher* 32(1994):415–417, and Timothy Slater, "Strategies for Using and Grading Portfolios," *Journal of Geoscience Education* 43, no. 3(1995): 216–220. For further information about laboratory grading, see Timothy F. Slater and Joseph M. Ryan, "Laboratory Performance Assessment," *The Physics Teacher* 31(May 1993): 306–308.

Grading is not nearly as complicated or time-consuming as one would imagine. Each portfolio (of three) counts for 20 percent of the course grade; laboratory work counts for 25 percent, and a final exam for 15 percent. Portfolios are graded as 0, 1, 2, or 3 on each learning objective described in their syllabus. A 0 score indicates no evidence: The information is not in the materials identified. A 1 score means the evidence is weak: The information is presented but not clearly, it is partially incorrect, or it reflects misunderstandings. A 2 score indicates adequate evidence: The information is presented with no errors or misunderstandings, but the information is dealt with at the literal, descriptive level, without integration across concepts. And a 3 score means strong evidence: The information is presented without errors or misunderstandings and implied in a comprehensive and integrated fashion.

Using this rubric, Slater finds interrater reliability to be high, and students themselves (in postcourse focus groups) report that they receive the grade they thought they deserved. (Sometimes they are more critical of themselves than their instructor.) They also report that test anxiety is considerably reduced. In addition, they pay more attention to class discussion, because they are relieved of vigorous notetaking and can concentrate on the physics of the situation, and they worry less in class about what each variable stands for, because they will be able to look that up later.

PARTIAL-CREDIT, "BLIND," WITHOUT LETTER GRADES, AND TAKING IMPROVEMENT INTO ACCOUNT

Daniel Stein, Department of Physics, University of Arizona, Tucson, AZ 85719; TEL: (520) 621-4190; FAX: (520) 621-4721; E-MAIL: dls@physics.arizona.edu.

Courses Taught

- Introduction to Mechanics (freshman level)
- Quantum Mechanics (advanced-undergraduate level)

Description of Examination Innovation

As chair of the physics department at University of Arizona, Dan Stein has taken a strong interest in improving physics students' academic experiences. As an introductory physics instructor, he was ranked "best professor ever" by one of the students we interviewed in Chapter 2. This student said that Stein "took the sting out of grading and made even a difficult subject like physics nonthreatening."

How does he do this?

First, Stein says he does not use trick questions and is generous with partial credit. For example, in multipart questions, a student's mistake in part A will not be held against the student in part B, provided he/she shows work, correctly done, in part B.

Second, Stein grades all his introductory physics students' exams (class size 60) anonymously. That is, students put their names on a cover sheet that the instructor does not look at when grading the exams. Stein insists on doing the grading himself, but "blind" grading guarantees that nothing that has transpired between student and instructor before the exam will go into any grading. This concern was voiced by many University of Arizona students we consulted about grading in our focus groups.

If Stein sees that a student (anonymous to him at the time) is having difficulty, he writes a note asking the student to come see him. (It is not unusual for Stein to spend two to three individual meetings a week with a student who is willing to put in the effort to master the material.) In addition, personally grading the exams allows Stein to gauge student understanding of the material and the appropriateness of the exam.

To that end, Stein does not assign letter grades on exams. He scores tests using points and then gives his class a range for a satisfactory score. "It's misleading to assign letter grades for each test because you can't make that determination of student progress until all the data are in," says Stein. He postpones using letter grades as long as possible. Students are kept informed, however, of their progress, with rough estimates of letter grades provided upon request.

Finally, when assigning final course grades, Stein will look at the overall trend in a student's performance, giving less weight to a low score on an early exam if the student has demonstrated improvement in understanding subsequently.

"I try to create an open, honest environment in my classes," says Stein. "I want them to think about how well they're understanding physics, not worry about their grades all the time."

Stein is well aware that some students are too intimidated by their professors to talk about their difficulties in the class. Stein thinks that, unfortunately, these problems will always exist, and all he and his department can do is keep encouraging students to use instructors as resources, and tell students instructors want to be helpful.

QUESTION CHALLENGE OPTION

Richard Stepp, Department of Physics, Humboldt State University, Arcata, CA 95521; TEL: (707) 826-5331; FAX: (707) 826-3279.

Courses Taught

- Introductory Meteorology
- Introductory Physics

Description of Examination Innovation

In two of his courses, introductory meteorology and introductory physics (both algebra-based), Richard Stepp has introduced a "question challenge" option, an idea he got from observing a jury selection. In his relatively large classes (60–120), for which he believes he has no choice but to give multiple-choice exams, Stepp allows students to "toss out" any four multiple-choice questions without being penalized for not answering them, so long as they replace each one with a short essay that does the following: (1) explains why they couldn't (or wouldn't) answer the question; and (2) demonstrates that they know the material. (But if a question is challenged, the student has lost the right to guess. *Any* answer on that question is erased by the grader.)

It typically takes the instructor no more than 2 hours to read 100 exam papers (Usually only a few students take advantage of the challenge option.) Stepp has found over time that most of the challenges point appropriately to poorly worded questions. More importantly, the new system "provides students with a *major* outlet," he reports. "Multiple-choice tests tend to reward best those who know exactly what the professor said, and may not know anything else. Thus, knowledge from another source that I didn't present can really confuse students, especially since questions (and answers) in introductory classes tend to involve greatly simplified versions of reality."

When taking a test, students may be confident of the point but not willing to challenge it until they are marked wrong, says Stepp. Thus he allows "late challenges," too. "Students like it best when they pin me logically to the wall."

The challenge "essays" are always brief, Stepp reports. Students don't have the time or energy on a timed test to produce more than a few lines. But he finds it to be a "major amelioration" of an otherwise inhumane format. Students love the option, says Stepp, though at first they believe it may be some kind of "trap" whereby Stepp will take their challenge personally. Occasionally, Stepp says he makes an error, and there is no right answer to a question. In that case, only those who challenge it get it right!

Stepp has used this option for the 10 years he has taught large classes.

MODIFIED PSI/MASTERY

Robert Svoboda, Department of Physics and Astronomy, Louisiana State University, Baton Rouge, LA 70803-4001; TEL: (504) 388-8695; FAX: (504) 388-1222; E-MAIL: phsvob@lsuvax.sncc.lsu.edu.

Courses Taught

- Astronomy 1101 (for nonmajors)

Description of Examination Innovation

For a two-semester course in astronomy for nonscience majors, Robert Svoboda has developed a "contract" system that offers students a "grade menu" from which to select how their grades will be calculated (see menu). The contract permits the student to tailor the course to his/her liking. Students who are poor at taking tests might decide to assign little weight to the final exam by doing a term paper instead. It is possible to skip the midterm entirely by doing a short report. Other students, who do well on tests, would make different selections, assigning all the weight to tests and quizzes and doing no reports, not even any homework. (Those who are "ultraserious" about grades, the instructor tells us, can do tests *and* reports, taking the better combination of grades.)

GRADE MENU

Appetizer

Percentage of Grade:
Choose One: □20% □30% □40%

Your Selection:
Choose One: □ In-Class Quizzes □ Homework

Main Course

Percentage of Grade:
Choose One: □30% □40% □50%

Choose One:
□ Final Exam Only (No Term Paper)
□ Term Paper (Final Exam to be worth only 10%)

Dessert

Percentage of Grade:
Choose One: □20% □30% □40%

Choose One:
□ Midterm Exam (no Short Report)
□ Short Report (no Midterm Exam)

Notice to Patrons: Make Sure Your Selections Add Up to 100%

Notes:

- You MUST select an item from all three catagories
- A new Grade Menu may be submitted at any time up to one day after the final
- If you select to do either a short report or term paper, make an appointment with me to get approval for the topic.

The system allows students to change their grade menu at any time up to the day *after* the final exam. Since accommodation to different learning and test-taking styles is embedded in the contract, all grading is absolute.

When asked about grading time, Svoboda explains that his short quizzes are half true–false and half short-answer questions that he grades in 30–40 minutes (with typically 50 students in a section). The homework is graded by a student grader. With 20 reports on average in a given section, they are by far the most time-consuming to grade. Overall Svoboda finds the system works well, and that once students get used to it, they like it, too. On course evaluations, the average student rating for the course is 9 on a scale of 1–10, with the basic science college average a 6. In fall 1995, Svoboda was named the best professor in physics and astronomy by the student government (out of 35 department faculty members) and was also given an award for the most outstanding professor at LSU from a local sorority. He believes his grading system may have had something to do with these awards.

TERM PAPERS, TERM PROJECTS

Marshall Thomsen, Department of Physics and Astronomy, Eastern Michigan University, Ypsilanti, MI 48197; TEL: (313) 487-8794; FAX: (313) 487-0989; E-MAIL: phy_thomsen@emuvax.emich.edu.

Courses Taught

- Introduction to Modern Physics (calculus-based)
- Kinetic Theory and Statistical Mechanics

Description of Examination Innovation

In the modern physics course, a one-term survey (for majors) of an assortment of topics in quantum mechanics, relativity, solid-state physics, nuclear physics, and particle physics, Marshall Thomsen assigns a term paper in which students are to write about one modern physics topic in detail, including references to journal articles. Although term papers of this nature are not unheard of in a physics class, they are the exception rather than the rule. Thomsen has positive feedback from most of his students on how valuable the library research has been, and at the same time, he has been able to gauge their ability to get into and learn about advanced topics in physics. Finally, Thomsen reports that the papers give him an opportunity to assess the students' ability to convey ideas in physics in formal writing (as opposed to using equations).

Thomsen also uses term projects in the advanced undergraduate/beginning graduate course in statistical mechanics. The small class size (5–10 students) makes this feasible. He selects a different problem for each student, so that the resulting work will reflect individual ability as opposed to a collaborative effort. (Collaboration is, however, encouraged on problem sets.) The problems are ones whose solutions are obscure enough that the students are unlikely to locate solutions in library research. Also, the instructor can tailor each problem to the student's background, providing, for instance, computationally intense problems to a student with programming skills. The pedagogical advantage of this innovation is that the students in Thomsen's physics courses have a chance to work through a complex, open-ended problem for which there are no answers in the back of the book.

A significant portion of the students enrolled in these two courses have been successful in pursuing careers in physics, primarily through graduate work or as high school physics teachers. Since many knew they were headed for graduate school, they valued getting an early taste of tackling challenging research problems.

Several colleagues in Thomsen's department use term projects, although previously, term papers were found only in the less technical courses such as History of Physics and Ethical Issues in Physics. Many department members recognize it is to the students' benefit to improve their writing skills within the context of the discipline.

POSTS "ADVANCE" ANSWERS TO PROBLEM SETS, "CRIB" SHEETS, OPTIONS FOR FINAL GRADE

Larry Viehland, Department of Science and Mathematics, Parks College of Saint Louise University, Cahokia, IL 62206; TEL: (618) 337-7575 Ext. 425; FAX: (618) 332-6802; E-MAIL: viehland@ions.slu.edu.

Courses Taught

- General Chemistry
- Thermodynamics
- Fortran
- College Algebra
- Physics (and labs for science courses)

Description of Examination Innovation

Larry Viehland engages in a number of examination and grading practices that, although they might not be new or unique, he finds useful. First, he posts worked-out solutions to problem assignments *before* they are due, so that students can get "expert" help when they get stuck. These problems are graded not on content or quality, only on whether they are completed. Second, he allows students to prepare one page of notes, which they can bring in to a test as a way of compensating those who do not have powerful alphanumeric calculators. It also has the indirect effect of getting students to learn as they prepare their own "cheat sheet."

Third, since he is more concerned with what a student knows at the end of a course than at other points along the way, and since students arrive in his lower-level courses with diverse backgrounds, he gives as a letter grade the higher of two scores, the average of all the test grades or the final exam grade. Fourth, he never gives multiple-choice exams. And fifth, each exam consists approximately of 50 percent number-based questions and 50 percent essay.

A downside to the first innovation—posting of problem solutions—is that some students simply copy Viehland's solutions. "They learn why this is a bad practice when they cannot work problems on the tests," he says. And some students complain that they spent too much time preparing a "cheat sheet" they never used on the exam. "After discussing this with them, it is clear, even to them, that the reason they didn't need it is *because* they prepared it," Viehland says. The other innovations listed earlier may make grading more time-consuming, but student feedback is generally good, and most students admit they're learning more. Other faculty colleagues use some of Viehland's techniques.

GRADING ON CONCEPTUAL UNDERSTANDING

Karl Vogler, Department of Physics and Astronomy, Valparaiso University, Neils Science Center, Valparaiso, IN 46383; TEL: (219) 464-5516; FAX: (219) 464-5489; E-MAIL: kvogler@exodus.valpo.edu.

Courses Taught

- The Essentials of Physics
- Introductory Astronomy

Description of Examination Innovation

With a wide diversity of students in his introductory physics classes (approximately 25 percent are freshman and sophomore nonscience majors, and the rest are upperclass premeds majoring in biology, chemistry, and related fields), Karl Vogler needed a flexible grading system that challenged advanced students without alienating nonscience majors. What he has come up with, and continues to develop, is an "across the board" method for grading conceptual understanding as well as "percent correct" on homework, quizzes, and exams. His system allows students to apply their strengths to a problem while learning to handle mathematical reasoning.

Final course grades are based on homework, quizzes, midterm, final, and scientific essay grades. Homework, worth 25 percent of the course grade, typically consists of two problem sets distributed weekly: one of easy to moderate problems, the other of moderate to difficult problems. Students with weaker math skills have the instructor's permission to turn in problems from only the first list, provided they attempt one problem from the "moderate to difficult" list. Students are encouraged to work collaboratively on the problems and to take advantage of weekly physics help sessions offered as part of the course. Before their final solutions are handed in, there is an in-class review, during which students take turns putting solutions on the board for the instructor and the rest of the class to discuss. The underlying concepts, quite as much as the steps toward the solutions are featured in those discussions. Final solutions are due the next class period and are only then graded by the instructor on a 1–5 point rating scale.

To earn full credit for homework, students must turn in at least 100 solutions of a "3" or better. (The less homework handed it, the lower the composite grade.) Students may work ahead if they wish, and special arrangements are made for students with weak math skills, including helping them find tutors and meeting outside of class with the instructor. Vogler finds that within 4–6 weeks, students with math difficulties no longer need extra help.

To achieve an interactive classroom environment, Vogler uses the in-class review as an opportunity for students to develop questions for class discussion about the material. The homework problems are distributed *before* the concepts have been discussed in class to groups of students who work together on a solution. One student from the group presents his/her group's solution to the class, and the instructor and other students discuss it. This collaborative "first pass" is not graded, although Vogler has considered doing so, especially because his nontraditional students say the group work is particularly helpful.

Part of the instructor's testing philosophy comes from his childhood experiences in athletics and music. In these arenas, the "instructor" is more of a coach than a lecturer. To continue the analogy, class and homework are "practice," quizzes are "practice games," and the midterm, final examination, and scientific essay are "real games," in which everything counts.

Another aspect of Vogler's grading innovation is that students are given significant partial credit for correctly setting up a problem and providing a detailed conceptual outline of the solution. They must draw a proper sketch and identify terms, forces, velocities, and so forth, as well as describe the concepts that apply to the problem and what mathematical relationships are needed to solve it.

"If the students can convince me that they could solve the problem, given a reasonable amount of time and no pressure, they can get about three-fourths of the credit for the problem," says Vogler, who wants to help the student who gets flustered when computing an answer to an exam problem. In this way, Vogler puts the non-math-oriented student on the same footing as the math-oriented student.

Vogler also allows students to select their final grade either from their class average or from their final examination alone—whichever is higher—provided they have fulfilled the homework portion of the course.

A downside to Vogler's grading system is that it confuses students who are so used to being graded on getting the "right answer." It takes them a while to grasp that they need to set up the problem correctly to get full credit. Others are slow to realize that incorrect answers are not necessarily penalized, because the homework they hand in is graded on some minimum number of problems that are correct. The student who does *more* than the minimum is advantaged in Vogler's system in two ways: first, in having a greater statistical probability of getting the minimum number of problems right; second, in having more practice.

Premed students, Vogler finds, are always trying to convert his 1–5 point homework scale into a percentage grade. But most students come to accept that group work and conceptual understanding are high priorities in Vogler's classroom. Significantly, one student from a prominent science preparatory school claimed he didn't need to do the homework and asked to be graded on exams alone. Vogler agreed, and the student flunked two semesters in a row. "He had a lot of good schemes for getting the right numerical answer using incorrect physics," said Vogler.

COMPUTER-BASED PAPERLESS TESTING

Andrew Wallace, Department of Physics, Angelo State University, San Angelo, TX 76909; TEL: (915) 942-2242; FAX: (915) 942-2188;
E-MAIL: Andy.Wallace@mailserv.angelo.edu.

Courses Taught

- Introduction to Physical Science
- Stellar Astronomy
- General Physics
- College Physics
- Engineering Dynamics
- Electrodynamics
- Applied Optics
- Solid State

Description of Examination Innovation

Andrew Wallace has been using computer-assisted instruction in the form of a paperless testing/homework system for 4 years in freshman service courses such as physical science. The large size of service courses (100+ students) and the scientific illiteracy of the entering freshman make a computer-assisted system a benefit in the classroom.

Students dump test administration and encrypted test-item files (only the instructor may alter the test-item bank) to a diskette at the beginning of the semester. Since each test is randomly pulled from a large test-item file, students can retake a test as many times as they wish until the course ends. Grades are logged to a central database, which saves the instructor time. It is the student's responsibility to log the high grade for each test.

Since it reduces test anxiety, students like the testing system. It also helps to build science literacy through repetition. Test items are limited to multiple-choice, true–false, matching, and short-answer questions. Students are required to submit a two-page summary of a current *Scientific American* article on their diskette at midterm. Students may also fill out the course evaluation form placed on their diskette anytime during the semester. Both the reading assignment and evaluation form are logged to the central database.

The downside of a computer-assisted testing/homework system is student cheating, says Wallace. Students will exchange information crucial to a test while at the computer. Some students spend more time trying to decrypt the test-item files than taking and logging tests. A computer testing center on campus will stop most student cheating.

Currently, one colleague is using a similar computer-assisted instruction and testing system in his finance course. Another colleague is using a different computer-assisted instruction system in his astronomy courses. The systems used in finance and physical science run on IBM-PC compatible machines in a DOS environment. The astronomy instruction system runs in a Windows environment. Interested persons may contact Wallace about these testing systems.

MULTIPLE-CHOICE, QUESTION CHALLENGE OPTION

Michael Zeilik, Department of Physics, University of New Mexico, Albuquerque, NM 87131; TEL: (505) 277-4442; FAX: (505) 277-1520;
E-MAIL: zeilik@chicoma.la.unm.edu.

Courses Taught

- Introduction to Astronomy

Description of Examination Innovation

In his introductory astronomy course at the University of New Mexico, Michael Zeilik, author of the textbook his students use,[14] doesn't like to use the test-bank materials provided by his own publisher. Instead, he has developed

[14] *Conceptual Astronomy* (New York: John Wiley, 1993).

another variant on multiple-choice that we have seen in modified forms elsewhere in this volume.

On his multiple-choice tests, Zeilik gives his students what he calls a "Challenge Opportunity," to identify an ambiguous question and challenge its range of answers. On each exam, students may select only one challenge opportunity and are admonished to "be sure" they have a substantial reason for the challenge, such as two equally possible answers. The "challenges" are to be done after the test is completed and, if accepted by the instructor, earn two extra points for the student. Challenging a question does not mean the student doesn't have to answer it as presented. It merely provides an additional opportunity to comment on it. Following is his description of his Question Challenge, which appears at the end of his exam:

> **Question Challenge**
>
> Questions are sometimes badly written, ambiguous, or have no answer that is clearly best. You are allowed to challenge *one* question by explaining briefly in the space below why you think the question is bad or inadequate. Valid and thoughtful challenges will be awarded two extra points.
>
> The question I challenge is number _______.
> The answer which I gave on the answer sheet is _______.
> In brief, this is why I challenge the question:
> (A whole page of space is left below for the challenge.)

About 20–25 percent of Zeilik's students will take the challenge option on any one exam. About half of this group earn only one point per challenge (their reasoning was not completely accurate), and about a fourth receive the full two points. Zeilik has noticed that student challenges over the last couple of years have gotten "less thoughtful and more trivial," suggesting to the instructor that students lack critical-thinking skills. Zeilik's colleagues think the method is a good one, though only a few have regularly adopted it for their classes.

MIXING SHORT AND LONG QUESTIONS, ALLOWING MAKEUP EXERCISES TO BE AVERAGED

John W. Zwart, Department of Chemistry, Physics, and Planetary Science, Dordt College, 498 Fourth Avenue NE, Sioux Center, IA 51250; TEL: (712) 722-6288; FAX: (712) 722-1185; E-MAIL: zwart@dordt.edu.

Courses Taught

- General Physics (algebra-based)
- Introductory Physics (calculus-based)
- Classical Mechanics
- Electromagnetic Theory

- Modern Physics

Description of Examination Innovation

John Zwart includes two types of questions in his introductory sequences: short questions and longer problems. For the short-question section, students have some options, such as doing five out of seven problems. The short questions involve a mix of computational problem solving and questions requiring written responses (e.g., "Is it possible to shoot a charged particle into a region where a magnetic field exists and *not* have a force exerted on the particle? If so, how? If not, why not?") The longer problems occasionally include essay questions focused on a specific set of concepts (e.g., "You saw a Cartesian diver in lecture demonstration in class. Explain how it works, citing relevant physical principles.")

"These questions help students see physics as being far more than simply 'the hunt for the right equation,'" says Zwart.

In addition to promptly grading, returning, and going over his tests, Zwart uses curve grading as appropriate. When scores fall below a certain expected range, suggesting the exam may have been too hard, he will curve the grades. Or, if the test uncovers a widespread misunderstanding of certain problems, he provides an opportunity to gain test points by doing take-home exercises and problems that address that particular misunderstanding. A scale that gives low scorers an incentive to work harder has also been devised by the instructor. The fraction correct on the additional work is multiplied by the number *wrong* on the test, and half this value is added to the exam score. Thus, if a student earned a 60/100 on the test and 90 percent of the points available on the makeup exercise, he/she would gain 40(0.9)/2, or 18 points. The grading system also rewards those who do well the first time around.

Zwart often provides the additional opportunity for makeup points after the first test in introductory physics. Rather than have his students feel "they are facing an uphill struggle for the rest of the course," Zwart provides them an opportunity for improvement to ease their anxiety. Zwart has learned from student responses that this now causes students to "shift from seeing me in an adversarial way; they recognize that I am more interested in their better understanding of the material than simply trying to get a good spread of grades." The course dropout/failure rate is down to a small percentage.

A few Dordt College faculty members have adopted this method (or a modified version) of dealing with low test scorers, and several of Zwart's colleagues also use conceptual essay questions in addition to quantitative problems on exams. Zwart uses essay-style questions on tests in upper-division courses as well.

CHAPTER 7

CONCLUSION

> There is nothing more difficult to take in hand, more perilous to conduct, or more uncertain in its success than the introduction of a new order of things.
>
> Niccolo Machiavelli, *The Prince*

The contributors to this volume are innovators. Each came to a conclusion after years of teaching that if his/her instruction was to succeed, some modification of standard in-class examinations was needed. In some cases, as we have seen, modest changes in the way questions were structured or in exam format or in grading practice were experimented with; in other cases, radical new exam ecology—group exams, open-book, take-home, laboratory assignments—were tried. Very often the examination innovation was part of a more general overhaul of curriculum and pedagogy. In other cases, the change in exam-taking procedure was intended to increase motivation and reduce anxiety and competitiveness. Sometimes, the innovation caught fire, and colleagues elsewhere in the department or in the college as a whole made comparable changes, or at least began to think about their own testing strategies. But, more often than not, at least according to our respondents, the innovation remains local; and in some fewer cases has been abandoned altogether because of the added workload on instructors already stressed by reduced budgets and staff, or because students balked at the extra or unfamiliar work.

What will it take to go from innovation to more permanent change? How will these early innovators and adopters persuade others that existing multiple-choice, machine-gradeable examinations in science are not conveying the right message as to what science is and what scientists do. Even faculty unfamiliar with the terms "reliability" and "validity" have strong feelings about fairness; they tend to believe that open-ended, open-book, group exams and alternate grading practices detract from the "objectivity" and "fairness" of assessment.

Yet, as thoughtful K–12 educators have long ago concluded, multiple choice doesn't eliminate subjectivity; it only displaces it onto the author of the questions. Moreover, reliability may get in the way of validity.

What will cause the faculty represented in this volume to see their recommendations followed elsewhere? And, most important of all, how are tomorrow's college educators in the sciences—today's graduate students and teaching assistants—going to be persuaded that reform of in-class examinations can and needs to take place?

These questions allow for no easy answers. But it is worth noting that Hellmut Fritzsche, a professor of physics at the University of Chicago, long ago added to his teaching assistants' training the requirement that they practice exam-creation and discuss under his supervision the issues of exam analysis and grading. And, Robert Yuan and his colleagues in biology and microbiology at the University of Maryland, College Park, are bringing their teaching assistants and graduate students (especially those intending to become college teachers) into their planning for examination reform.

College and university administrators play a critical role in all of this. If inadequate staffing stands in the way of widespread testing reform, modest infusions of additional resources could make a profound difference; certainly, it would send the message that administrators acknowledge the importance of in-class testing in the sciences.

This volume has focused on innovation and change in lower-division courses in college-level science. In Part 2, we introduce comparable instructor-generated entries designed to improve the testing environment in upper-division courses.

APPENDIX

Sheila Tobias
Jaqueline Raphael
P.O. Box 43758
Tucson, Arizona 85733-43758

To: Friends and Colleagues
From: Sheila Tobias and Jacqueline Raphael, co-authors
Subject: Your Contribution to a Manual on Testing

The California State University System has recently inaugurated a unique university press, one that will serve the teaching faculty in post-secondary institutions with useful how-to manuals and handbooks. The first of these, *Computers in the Classroom*, was published in 1990 and has been distributed at near-cost to postsecondary educators in the California State University System and elsewhere. The second, now in preparation, will be a compendium and analysis of new ideas in testing practice for the teaching of college science entitled: In-class Examinations: New Theory, New Practice for the Teaching and Assessment of College-level Science

Why This Project?

It has been said that examinations are the "latent curriculum" that, more than what faculty *say* they want students to learn, drives student behavior. Yet, in most undergraduate science courses, the same testing practices have been used for years. This, despite much criticism. Reasons for not making change range from habit, the convenience of textbook-generated exams, the cost of hand-grading as against machine-grading, and so on. Still, the issue ought not be ignored. This manual will serve to put it on the teacher-innovator's agenda and serve to support that teacher–innovator when he or she approaches a dean or department chair for financial or other support.

What Is Planned

This 100-page manual will consist of three parts: Part 1, a 30-page analysis of in-class examinations as currently employed in college-level science courses; Part 2, a 60-page collection (and commentary) of innovative and exemplary testing practice at the undergraduate level in science (possibly mathematics); and Part 3, analysis, conclusion, and annotated bibliography, written materials, computer-related materials, contacts, and other resources. In addition, a list of funders with an interest in testing will be appended to the volume.

How You Can Become a Contributor

If you have designed and used any testing innovations, we would like you to contribute to this project. Do not, incidentally, let the word *innovation* intimidate you. We define it as a unique or experimental examination practice—in exam design, format, style, or grading—that enhances the quality of learning in your classroom. Here are some categories—by no means an exhaustive list—that have already been identified as areas for innovation:

1. Exam design: content of test items, test item construction
2. Exam format: verbal, pictorial, quantitative, open or close-ended, multiple choice, etc.
3. Exam environment: individual, group, in-class, take-home
4. Exam grading practices: pass/fail, curve/absolute/resurrection or other point- compensation schemes, "contracts"

Initially, we will simply be collecting contributors' names and a brief description of their innovation(s). Later, the authors will get back to contributors for more detail in a telephone interview and may ask for additional written commentary on how the innovation was developed, how it was received (by students and colleagues), whether it is continuing, and, if not, why not.

All Contributors Will Have the Opportunity to Edit and Approve Any Description of Their Innovations.

You can help us by completing the enclosed brief summary sheet, which will give the authors all the information they need to start the collection and commentary process. There's room on that sheet for the names and addresses of other colleagues whose work deserves to be included in this book. If you need more information about this project, please feel free to call Sheila Tobias or Jacqueline Raphael.

Please answer on this or separate piece of paper and return the summary sheets to

Sheila Tobias
P.O. Box 43758
Tucson, Arizona 85733-43758

Assessment Innovation

Name and address (please include phone and fax)

Course(s) taught:

Describe what new examination form and/or practice you have been employing and for how long you've used it using the remainder of this page and the reverse side as needed.

Please list names and addresses of colleagues who might have something to contribute to this volume.